Ethics of new reproductive
technologies

Ethics of New Reproductive Technologies

Teaching Ethics: Material for Practitioner Education (TEMPE)

General Editor: Donna Dickenson, John Ferguson Professor at the University of Birmingham and Director of the Centre for the Study of Global Ethics

These three books introduce key areas in current medical ethics to readers with no previous knowledge in the field. Structured around a variety of guided activities and real-life cases, the books look respectively at the implications of the possible abuses of new reproductive technologies, "new genetics", and protection of research subjects in an increasingly global environment of research trials. Both authors and cases represent a wide range of European backgrounds and professional disciplines, including medicine, bioethics, law, sociology, theology, and philosophy. A video further enhances the value of these workbooks.

Volume 1

Ethics of New Reproductive Technologies: Cases and Questions

Dolores Dooley, Joan McCarthy, Tina Garanis-Papadatos, Panagiota Dalla-Vorgia +

Volume 2

Ethics and Genetics: A Workbook for Practitioners and Students

Guido de Wert, Ruud ter Meulen, Roberto Mordacci, Mariachiara Tallacchini

Volume 3

Issues in Medical Research Ethics

Jürgen Boomgaarden, Pekka Louhiala, Urban Wiesing

Ethics of New Reproductive Technologies

Cases and Questions

Dolores Dooley
Joan McCarthy
Tina Garanis-Papadatos
Panagiota Dalla-Vorgia

Berghahn Books
New York • Oxford

First published in 2003 by **Berghahn Books**
www.BerghahnBooks.com

Library of Congress Cataloging-in-Publication Data
Ethics of new reproductive technologies : cases and questions /
Dolores Dooley ... [et al.]
    p. cm -- (Teaching Ethics: Material for Practitioner Education; vol 1)
  Includes bibliographical references and index.
  ISBN 1-57181-599-6 (hardback : alk. paper)
    1. Human reproductive technology--Moral and ethical aspects. 2.
Infertility--Treatment--Moral and ethical aspects. I. Dooley, Dolores. II
Teaching ethics (New York) : vol. 1.

  RG133.5 .E848 2003
  176--dc21

British Library Cataloguing in Publication Data

A catalogue record for this book is available from the British Library
Printed in the United States on acid-free paper

ISBN 1-57181-599-6 hardback

Contents

In memory of Panagiota Dalla-Vorgia

Acknowledgements

We owe a debt of gratitude to a great many people who contributed in different ways to this text. *The Ethics of New Reproductive Technologies: Cases and Questions* is the result of a partnership between Imperial College, London, the University of Athens Medical School, the National School of Public Health in Athens and the National University of Ireland at Cork, Ireland. The cases, commentaries and articles, which form the core of this book, have been gathered from all over the European Union, the United States and Australia. Many of them were contributed by participants at a series of workshops held in Cork and Athens, as part of the Teaching Ethics: Materials for Practitioner Education (TEMPE) project which was funded by the European Commission's BIOMED III programme. Our sincere thanks go to each of these participants: Ron Berghmans, Donna Dickenson, Wybo-Jan Dondorp, Mary Donnelly, Sigrid Graumann, Fiona MacCallum, Deirdre Madden, Linda Nielsen, Francoise Shenfield, Vassilis Tarlatzis, Hugh Whittall and Heather Widdows.

Our thanks go too to those participants who commented on the first draft of the text in an Assessment Workshop in Cork in 2001. Their comments and criticisms, which have been incorporated into the final text, have greatly improved its clarity and precision. Participants at the Assessment Workshop were Carole Barry Kinsella, Colin Bradley, Helen Browne, Denis Cusack, Vittorio Bufacchi, Gerardine Fitzpatrick, Nora Geary, Caroline Harrison, Richard Hull, Geraldine McCarthy, Orla McDonnell, Nora Mansell-Quirke, Angela O'Connell, Rhona O'Connell, Patricia O'Dwyer, Georgios Papagounos, Zaira Papaligoura, Matthew Ratcliffe, Tony Ryan and Elena Tsourdi.

First drafts of each of the chapters were sent to critical readers in several countries whose suggestions have greatly contributed to the success of the book. The critical readers were Nikola Biller-Andorno, Claus-Dieter Middel, Sakari Karajalainen, Ritva Halila, Lucy Frith, Massimo Reichlin, Giuseppe Tavormina, Hugh Whittall, Michael Parker and Inez de Beaufort.

The authors warmly acknowledge the contribution of the series editor, Professor Donna Dickenson, Director of the Centre for the Study of Global Ethics, University of Birmingham. As lead coordinator of the TEMPE project based in Imperial College, London, Professor Dickenson provided immeasurable assistance and advice over the two-year period of its duration.

Finally, we would like to make special mention of Professor Panagiota Dalla-Vorgia, University of Athens, who passed away right after the completion of this text. Her participation proved invaluable to this project and she will be remembered by all those who worked with her, not only as a highly competent and respected professional, but, as a warm and generous colleague as well.

Introduction

Ethical reflection is a challenge for us as human beings, as students and as professionals who want to delve deeper into reasons why we cherish specific values and ideals. Specifically, the advent of 'new reproductive technologies' (NRTs), sometimes called 'assisted reproductive technologies' (ARTs), has faced us with increasingly complex ethical situations to unravel and difficult choices to make. These new technologies encompass three different kinds of procedures which have been described by Kymlicka in the following way.

New Reproductive Technologies (NRTs) involve:

- procedures intended to assist individuals or couples to conceive a child (e.g., artificial **insemination, in vitro fertilisation (IVF)**, and surrogate childbearing);

- procedures intended to assess and promote the health of the **embryo** or foetus after conception (e.g., prenatal diagnosis, preimplantation genetic diagnosis (PGD), sex selection, embryo therapy or foetal surgery);

- the use of human gametes, embryos or foetuses in research (e.g., embryo experimentation, or the use of foetal tissue transplants in the treatment of Parkinson's disease). (Kymlicka 1996: 262)

So, NRTs include procedures that assist conception and generally assess and promote the health of the embryo. They also involve research aimed at preventing or curing infertility as well as other therapeutic goals. Let's pause for a moment and try to make more concrete a specific range of choices available with NRTs. Briefly, given the developments in embryological research in the last twenty years, it is now possible to:

retrieve sperm and ova (gametes) from men and women

store the gametes through a freezing process (**cryopreservation**)

fertilise sperm and ova in a laboratory test tube

With successful fertilisation, embryos may:

be implanted in a woman to achieve a pregnancy

be implanted in a surrogate woman who carries a pregnancy for

someone else

be stored through freezing

to be used for a future implantation in gamete providers or

to be donated to another infertile individual or couple

to be used for research purposes

to be used for deriving **stem cells** for therapy or research

to be genetically tested for 'quality or health' of embryos

Or embryos may be destroyed if none of the above options is desired.

Activity:

Clinicians and researchers will be involved with making decisions on all or most of these options. Professionals in general practice will be seeing individuals or couples who deeply desire to have a family. These individuals and couples are increasingly doing a great deal of homework to find out how they can have a child who will be spared a hereditary disease or other disability. They are coming to family doctors and nurses to hear what they recommend and what they think of the risks, benefits and likely outcomes. If you look back at the list of options given above: What were your first responses to the items on the list? Did some options stop you up more than others? Which were they? Do you know why you paused at some more than others?

Aims and objectives

This workbook is not intended to cover all of the possible topics that one might include under the umbrella title of *The Ethics of New Reproductive Technologies*. This is because, as indicated in the foreword, the workbook is the outcome of an

European Union-wide collaborative process. It has been prepared using papers, case studies and commentaries from ethicists, lawyers, policy-makers, clinical practitioners, educationalists and other experts working in Europe. In the course of this process, attention has focused on what are perceived to be the 'burning issues' in different European countries and it is on the basis of this that the topics in the workbook have been chosen.

We view the workbook as a flexible educational resource for the medical and nursing curriculum and for post-graduate training or distance learning programmes. We encourage both learners and teachers to use the materials in a way that best suits their requirements. For some, this might mean working through each section of the workbook, but others might choose to focus on a particular case study and its related activities and use it in conjunction with other materials. We also suggest that many of the activities in this workbook are particularly useful as group exercises and as a basis for group discussion of the various ethical issues at stake. Where the analyses and discussions introduce new or unfamiliar terms (highlighted in bold), the reader can refer to the Glossary at the back of the book. There is also an Appendix of tables which indicate the legislative and regulatory provisions of each of the EU countries with regard to NRTs.

The learning objectives of *The Ethics of New Reproductive Technologies* are outlined below. You might review these objectives from time to time as you work through each section, but at the end of this workbook you should be able to:

- define and analyse a range of ethical terms that are relevant to reproductive decision-making such as reprogenetics, responsibility and quality of life
- tease out the values, ideals and principles that are 'at risk' in the case studies
- weigh the arguments for and against maintaining donor anonymity and secrecy within the family in relation donor insemination
- critically discuss the impact of surrogacy arrangements on surrogate mothers, commissioning couples and the children involved
- analyse different views on the status of the embryo and consider the implications of these views for accepting or

rejecting a ban on embryo research, specifically, on embryonic stem cell research
- explain the way in which notions of 'property' and 'consent'
  are deployed in the common law tradition and evaluate three
  possible approaches to the ownership of embryonic and
  foetal tissue
- evaluate the implications of the conflicting professional
  roles of the IVF clinician who is both *doctor* and *researcher*
- assess the impact of human egg donation on the health, well-
  being and interests of women and couples
- analyse the notion of a 'right to procreate' in relation to
  access to IVF services and critically discuss the potential
  conflict between the parents' autonomy and the child's
  welfare
- consider the relationship between individual liberty and the
  public good and assess the personal, social and political
  consequences of the continuing use of genetic testing for
  reproduction

Why do we need ethics?

You may notice that none of the learning objectives listed
above can be met through rote learning of a list of facts. Rather,
what is required is careful thought, argument and appreciation
that there is seldom a single, simple, universal solution to any
of the ethical dilemmas posed in this workbook. It is precisely
the complex nature of the case studies that we have provided
which demands a particular kind of attention and reflection.
This is because ethical claims are not like factual claims which
can be proven or disproven on the basis of appeal to evidence
or experiment. On the other hand, we cannot simply justify
them on the basis of appeal to personal preference or attitude.
In clinical practice, multidisciplinary teams need to make decisions while at an organisational and social level, ethics
committees, professional groups and government departments
need to decide on guidelines. How can this be done if everyone's opinion is equally good? In short, there are no simple
solutions to the ethical dilemmas that we have.

One way of addressing the complexity of moral decision-
making and disagreement is to take a set of principles such as
those offered by Tom Beauchamp and James Childress (2001)

as a basis for medical ethics. These can also be considered as values or ideals used to facilitate decision-making. They don't deliver 'automatic' and unproblematic decisions but they contribute to a reflective process of assessing features of ethical situations such as the cases that follow in this workbook. If one regularly used these principles in decision-making, one would be following a **principlist approach in ethics.**

Respect for autonomy (self-determination): To respect an autonomous person is, at minimum, to acknowledge that person's right to hold views, to make choices, and to take actions based on personal values and beliefs.

Nonmaleficence: Inspired by the Hippocratic Oath, nonmaleficence states the obligation not to inflict evil or harm.

Beneficence: Refers to the moral obligation to act for the benefit of others. Beneficence on this understanding refers to the character trait or virtue of being disposed to act for the benefit of others. A principle of beneficence asserts an obligation to help others further their important and legitimate interests.

Justice: Different philosophers use the terms 'fairness', 'entitlement' (that to which one is entitled) and 'desert' (what is deserved) to explain justice. These accounts, understand justice as equitable, fair and appropriate treatment in light of what is due or owed to persons. So, an injustice involves a wrongful act or omission that denies people benefits (health care) to which they have a right or fails to distribute burdens (taxes) fairly.

Activity:

What you do think of these as starters for moral reflection? You may already see that this list of four principles doesn't provide a smooth ethical journey since, in some cases, there may be conflicts due to:

- clear disagreement on the interpretation of these principles, or
- disagreement about how they should accurately be applied to a particular case, or
- disagreement about how to rank them or prioritise them.

Alternatively, viewing ethics as, primarily, a **communicative activity** extends the process of ethical decision-making beyond a simple appeal to moral principles. Consider the following passages by Dieter Birnbacher:

> A conception of ethics as a theory has tended to predominate in the tradition of philosophy. Its task was the theoretical clarification of moral concepts, the study of moral arguments and the development of maximally coherent and well-founded sets of moral principles. Ethics in this sense was academic work done in writing books, giving lectures and holding seminars.
>
> Ever since the time of the Sophists and of Socrates, however, there has also existed a rival conception of ethics according to which ethics is practice rather than theory, more analogous to art than to science. According to this conception, ethics (and philosophy generally) is an activity rather than a doctrine, where 'activity' means an essentially communicative activity of problem identification, deliberation and problem solving. Though making use of methods similar to those of theoretical ethics and requiring similar skills (in fact, some more), this kind of doing ethics is quite different in its performance aspects. Ethics as theory is mainly monologue, ethics as practice mainly dialogue. Ethics as theory deals mainly with intellectual problems and intra-disciplinary controversies, whereas ethics as practice deals mainly with real-life problems and extra-disciplinary controversies. Ethics as theory deals mainly with potential cases, ethics as practice mainly with real cases... The idea of medical ethics is a truly practical discipline. It derives from the Socratic idea that ethics (and philosophy generally) is an activity and an activity that is, in principle, open to everyone prepared to subject himself to the discipline of controlled dialogue. (Birnbacher 2001)

Activity:

Take a few moments to consider what might be involved in Birnbacher's notion of 'controlled dialogue'. Make a list of what you think the key elements of ethical dialogue should be.

We offer the following suggestions. As you read through the list, compare it with your own adding any of those that are missing to your list.

Ethical dialogue should include:
1) The identification, definition and analysis of the issues at stake.

2) The adequate expression and articulation of differing views.
3) The provision of plausible, factually correct and relevant arguments in support of differing views.
4) The effort to understand others' views and arguments: the background assumptions which inform them and the financial, personal, legal or institutional constraints which guide them.
5) The openness to revise one's views in the light of 4.
6) The effort to secure either consensus or innovative and imaginative solutions to the ethical dilemmas at stake.

In sum, the principlist and communicative approaches to ethics offer two ways in which to address ethical issues and dilemmas that do not involve an appeal to subjectivist and relativist attitudes (as defined below). Importantly, they are not incompatible with one another. Principles must be applied in practice and practice involves dialogue which requires certain standards such as those outlined above. Alternatively, when in practice, we defend our positions to one another, we often appeal to such principles as Beauchamp and Childress (2001) have identified.

As you will see progressing through each section of this workbook, we have included both of these approaches (among others) to ethical decision-making in a way which, we hope, you will find useful. From now on, it's up to you.

Briefly ....

Ethics or morality is concerned with what we ought to do, how we ought to live and act. Ethical or moral claims are claims about what it is to live a good life and often contain such notions such as *responsibility, right, good, duty,*etc.

An **ethical theory** usually consists of a small number of basic moral principles and often makes claims about what it is to be human or what it means to be moral. It is intended as a mechanism to solve moral problems.

A **moral principle** is a basic claim defining what makes actions right or wrong. Note the difference and the relationship between principles and values. A principle is a basic *rule* that protects a fundamental *value*, e.g.,

Principle: treat people equally. *Value* (protected by the principle): equality.

An **ethical dilemma** arises when two or more values are in conflict, e.g., a dilemma might be how do I provide adequate care for a patient and at the same time protect his or her autonomy?

Those who take a **subjectivist** position with regard to ethical claims argue that morality is a matter of attitude, of personal preference where conflicting positions need no defence.

Those who take a culturally **relativist** position with regard to ethical claims argue that whatever the majority (or perceived majority) in any society *thinks* is right or wrong determines the moral status of any action as it applies in that society.

Those who take an **objectivist** position with regard to ethical claims argue that there are certain (universal) moral truths that would remain true whatever anyone or everyone thought or desired.

Notice that neither the principlist nor the communicative approach to ethics claims to offer objective truths. But nor are they simply subjectivist or relativist. The aim of both approaches, but particularly that of communicative ethics, is not to prescribe particular outcomes but to ensure that the process of ethical decision-making meets certain standards.

- 1 -

Rethinking Reproductive Responsibility

Objectives

At the end of this chapter you should be able to:

- explain the term reprogenetics
- clarify components of responsibility: professional and parental
- discuss the way in which judgments about 'quality of life' are both medical and value laden
- provide reasons for your decisions on cases presented and reasons why someone else might object to your view

Reprogenetics

What is beginning to catch the attention of the public is the unprecedented power and the difficult choices that result when current technologies for assisted reproduction and genetic knowledge are brought together. On the one hand, an embryo can be grown *in vitro* and its sex can be readily identified. On the other, the mapping of the human genome is providing increasingly specific understanding of which genes are implicated in conditions such as Huntington's disease, cystic fibrosis, Duchenne muscular dystrophy, breast cancer, diabetes and deafness. A new concept and relatively new term available for the joining of these two medical technologies is *reprogenetics.* Reprogenetics and the combined technological potential it represents symbolises a seismic shift from a situation where birth, death, disease and disability happen by chance to a situation where choices allow us to avoid or modify the seeming 'inevitable' nature of our human world.

Dooley (2000) argues that new reproductive options may instigate major revisions in our thinking about ourselves and specifically, about how we understand the notion of responsibility. This is because the chance-choice distinction touches deeply into the core of our identity as moral agents. We use the chance-choice distinction in our assignments of responsibility for situations as well as in our assessments of pride, including pride in what nature has given us. She poses the following questions:

> what happens if it becomes commonplace that reproduction, character traits and human attributes are not inexorable results of a chance nature? What happens if such human attributes, traits and details of physical features move to the arena of choice? If they were a matter for choice, would these possibilities extend the range of our responsibility? (Dooley 2000)

Dooley concludes that infertility has always been seen as an unfortunate happening that we had to live with because of the chance distribution of genes or God's will or fate. Moreover, she claims that there is a sense of 'moral dislocation' if new knowledge creates new opportunities for choices about re-creating and redesigning who we will be and what capacities we will have.

Activity:

Before we continue, stop and ask: What areas of living are now opportunities for choice that were previously accepted as chance, as nature's designs for us or God's unknown will? What about simple illnesses such as TB, pneumonia, chicken pox, diabetes, high blood pressure? Just seventy-five years ago we were subject to the vagaries of suffering from these illnesses and now medications and vaccines are available. With antibiotics we no longer face death from contracting bacterial pneumonia. Consider also therapeutic and technological options for prolonging life. The minimising of chance in our life means we need to assess difficult choice options. Can you name other momentous shifts from living with chance to pursuing choice? The following questions may seem a daunting list for reflection. However, the possibility of moral vertigo might help us spell out some questions and probable projections.

1. Will reprogenetics reform the kind of obligations parents think they have towards their yet-to-be children?

2. Will parents come to be expected to use antenatal genetic testing or pre-implantation genetic diagnosis to ensure serious genetic diseases are avoided? To ensure that some 'preferred traits' are better assured?

3. Will future progeny blame parents if they could have taken measures to have them 'different' or 'healthier' but didn't make that choice?

'Simple' IVF

The world's first test-tube baby, Louise Brown, was born in England in 1978. In less than twenty-five years, the possibilities and options for reproductive choice have hugely increased creating a comparable demand for ethical reflection, discussion and decision-making. The multiple options now offered through fertility clinics are largely a development of the potential in the technology of *in vitro fertilisation* (IVF). Consider the situation of infertility in the following case and use it as a starting point. Begin to formulate questions about some of your values and ideals that you think are: fostered, diminished or under serious threat in the narrative of the case.

The Case of Alice and Martin

Alice Goodwin was told she had a massive infection in the fallopian tubes and surrounding area which seemed resistant to treatment. As a consequence, both fallopian tubes had to be removed. So at the age of nineteen, Alice felt depressed at the thought that she might never have children who were her own. Alice married four years later and the worry about being childless became acute.

One of Alice's friends was a midwife for the health authority in the district where Alice lived. She told Alice, and her husband Martin, about a fertility clinic only 40 miles away that had been counselling couples for possible IVF. A visit to their family doctor confirmed that they were ideal candidates for IVF. They could provide the sperm and ova and, if successful, the child would be biologically their own offspring. This realisation of 'their own biological child' was so important partly because that is how they understood 'being a parent' and also because they both had a strict Catholic upbringing and knew how their Church felt about 'unnatural procreation'. However, much discussion together cleared their conscience. As they put it: 'We considered that we had a mechanical breakdown in our reproductive equipment and a mechanical remedy by the IVF team was necessary!'

Martin and Alice felt encouraged. It was October 1980, early days with IVF, when Alice and her husband, Martin, had their first attempt at IVF at the recommended fertility centre. After two years without success they had their fourth attempt at fertilisation in vitro. Two fertilised eggs were transferred to Alice's womb and the emotional anxiety of waiting to see the outcome became more acute. She felt the clinicians had given no understanding of the stress that the hormone treatments for ovulation and the waiting game caused. One of the twins died at eight weeks but the eventual birth of their daughter brought happiness and profound relief. They knew that someday the technology would enable them to store/freeze extra embryos for future pregnancies. But the time was not yet. Alice and Martin told the team in the IVF centre: 'After sorting out some initial reluctance coming from our religious upbringing, we feel very good about our decision; we don't see any ethical or moral problems with what we have done. We only hope for the sake of other couples that IVF programmes flourish.'

Activity:

(a) Do you think that there are any ethical or moral problems in this scenario? What about requirements for Alice and Martin to be able to give informed consent? Were potential risks made clear? Was the success/failure rate explained?

(b) Try to give reasons if you differ with Alice and Martin's choice. It will help if you try to work towards being as specific in your reason-giving as possible. For example: 'I think they jumped too quickly into IVF rather than considering the option of adoption; adoptions would have been much more unselfish and helpful to already existing children who need homes.'

(c) Can you explain the objections to IVF, mentioned as coming from the Roman Catholic tradition? How do you evaluate these objections? Do you know of other religious positions on IVF? Remember that religion, like culture, like family context, may well be a basis from which people begin their moral development and acquire moral values.

(d) Is IVF freely available in your country for both public and private patients? Is it well advertised? Explain why IVF is not readily available if that is the situation? Are the reasons economic, political, ethical?

This initial case of Alice and Martin is an example of a 'simple case' of IVF. Many people would think this is a misnomer. They would argue that there's nothing simple about IVF! In terms of the process which usually involves high dose hormonal treatment to achieve multiple eggs and the psychological stress of waiting and wondering the outcome, IVF is anything but 'simple'. But 'simple' here means only that the eggs and sperm (gametes) are given by the 'intentional couple' – i.e., the couple who intends to carry and rear the resulting offspring. Some of the chapters that follow will involve reflection and questioning on the values and ideals at issue in more 'complex' options of assisted procreation. So starting here may facilitate moves to greater complexity.

Now let's look at a case of 'complex' IVF where IVF is combined with genetic testing of embryos at pre-implantation stage. What has the shift from chance to choice meant for our actors in this life narrative?

The Case of Linda and Philip

Linda and Philip are both deaf. Linda is infertile and the couple have been accepted for IVF. Once on the programme, they were offered Preimplantation Genetic Diagnosis (PGD) by a well-meaning clinician who assumed that they would not want any of their children to be deaf. He is shocked when they steadfastly insist that out of their nine embryos, those with congenital deafness be implanted first – along with any one of the other unaffected embryos. The remaining embryos should be frozen for later use.

They justify their decision by arguing that their quality of life is better than that of the hearing. As far as they are concerned, giving preference to the affected embryo is giving preference to the one which will have the best quality of life. They are very concerned that any hearing child they have will be an 'outsider' – part neither of the deaf nor of the hearing community at least for the first five or so years of his/her life. (Draper and Chadwick 1999: 116)

Note: See Appendix 3 for information on EU legislation in relation to the provision of PGD.

Activity:

(a) Linda and Philip stunned the clinician by their reasons for wanting to have PGD. Do you share that stunned response? Try to detail some of the reasons why the clinician might be shocked at this preference for a deaf child. Make a list of the attitudes the doctor might have toward deafness and/or deaf people and any values he might hold that generated his reaction. Did this decision for a deaf child go contrary to the clinician's sense of his professional responsibility to the well-being of the child?

(b) If a couple decided to adopt a deaf child, we might spontaneously consider them altruistic, generous and very compassionate. But if a couple decides to use available reproductive technology to try and achieve a deaf child, are we as confident that we would use the same adjectives for the couple? Discuss this among your friends, partner or colleagues. Is there disagreement? Why?

(c) Begin a list of points on this case to gradually develop your idea of *responsibility*. But first consider Tina Garanis-Papadatos' (2000a) commentary on the case of Linda and Philip.

Commentary on Linda and Philip

Tina Garanis-Papadatos

This case brings to the surface an issue which will become more common as genetic technology moves forward: the desire and the subsequent possibility to have a child carrying specific genetic characteristics. Although it might be considered closer to human nature to seek the birth of a child with better health, more beauty or enhanced intellect, the opposite has also been happening, causing serious ethical dilemmas: namely cases where people who have a certain disability wish to have children with the same traits. This attitude has been found to be more common in two cases: achondroplasia (dwarfism) and deafness. Preimplantation genetic diagnosis can make the wish of these people come true. It can give them the possibility to elect the implantation of an embryo which carries the 'desirable genes'. Although this decision of Philip and Linda takes the attending physician by surprise and sounds completely irrational to him, the couple's argument that their quality of life is better than that of the hearing reflects one prevailing attitude in the Deaf community. The desire to have a deaf child with the help of modern genetic technologies is not purely an ethical problem about genetics.

Activity:

You probably noticed in Garanis-Papadatos' commentary here that she observes the convention of capitalising 'Deaf' to refer to the cultural group of sign language users rather than the lower case, 'deaf' referring to the physical phenomenon of not hearing. Did you understand what Garanis-Papadatos means when she says that the desire and choice to have a Deaf child is not a purely ethical problem about genetics? What kind of problem do you think it is? She emphasises that there is an important judgment call here about 'quality of life'. What would you include in the idea of 'quality of life'? Let's take a moment to think about that.

Discussion

The idea of 'quality of life' will surface at many points throughout the text: we might talk about quality of life as it pertains to a child, a sick person, an elderly dementia patient, a paraplegic, a neo-nate, a dying cancer patient, etc. We can also talk about the 'quality of life in a given society' which draws in a range of considerations that will be the subject of our final section of this text. But for now, see what you think of the ideas about 'quality of life' in the following discussion.

'Quality of life' is a term that is widespread in health care ethics but we often don't stop to consider what we mean by it. Some people react vigorously to any suggestion that 'quality of life' judgements are made in medicine. This reaction arises often because it is thought that judging 'quality of life' implies some possibility of a quantitative measure of value on lives. But when Philip and Linda say that they believe their Deaf offspring would have a better 'quality of life' than a hearing child, do you think they are somehow wrong to make that judgment? Parents do make such judgments all the time when deciding whether or not to send their children to school x rather than y or when they decide to have their child receive 3 in 1 vaccination, etc.

'Quality of life' is difficult to define and interpret because there are many different ways of thinking about it. The fact that no definition is universally acceptable indicates that we need to look more closely at what varied elements inform an idea of 'quality of life'. A 'good quality of life' may mean different things to different people and, with the passage of time, the things that matter, things we most care about change. Compare, for example, what was judged to be 'a good quality of life' among European working classes of the fifteenth century with judgements that might be made today in different parts of the world. The provisions and opportunities in life change our hopes and aspirations about what is possible and what is desirable. Such changes greatly alter our understanding of a desirable 'quality of life'.

However, it is broadly accepted that quality of life is not concerned simply with the presence or absence of physical illness. It is not simply referring to all senses and limbs' physically functioning. Many people enjoy that but they yet consider that they do not enjoy a quality of life. So, many other elements might be included under the umbrella concept of 'quality of life':
• possibilities of variety of pleasurable experiences,
• meaningful relationships, (intimacies and/or acquaintances)
• purpose for getting up in the morning,
• spiritual traditions in one's life
• emotional and psychological supports in daily living
• availability of opportunities to grow and improve one's human potential etc.
• commitments to important causes
• values of our culture or sub-culture

Returning to the narrative of Linda and Philip and the stunned clinician: if the clinician disagrees with the couple about the possible 'quality of life' of the future child, consider what the basis for the disagreement might be. One area of disagreement might be that the clinician has a strong medical model of understanding of 'quality of life' where he means specific features of a 'healthy, whole, seeing, hearing, walking, cognitively functioning infant, etc.' He may also never have had any acquaintance with any Deaf patients nor seen the culture of Deaf people up close. However, a medical understanding of 'quality of life' here couldn't embrace the experiential, cultural and value inputs at the basis of the couple's preference. So it is very doubtful that objective medical factors can carry the weight of judging quality of life in the case of Linda and Philip.

John Harris speaks to this question of selecting for deafness on the grounds of an anticipated superior quality of life. Harris' points may well be applied to other such choices. In making the clinical decision, Harris argues that he would not endorse overriding the parents' procreative choice in these circumstances. Further, the doctor would be acting wrongly if he overrode the parents' choice. But, even if the parents are entitled to choose which embryos to implant, it does not follow from this, that they are acting rightly if they implant the deaf embryos. Harris defines disability as a 'physical or mental condition that we have a strong [rational] preference not to be in and it is, more importantly, a condition which is, in some sense a 'harmed condition' (Harris 2000: 96–7). He asks us to imagine a case where a cure for congenital deafness is discovered that is risk free with no side effects and poses the following questions:

> Would the parents, in this case, be right to withhold this cure for deafness from their child? Would the child have any legitimate complaint if they did not remove the deafness? Could this child say to its parents: 'I could have enjoyed Mozart and Beethoven and dance music and the sound of the wind in the trees and the waves on the shore, I could have heard the beauty of the spoken word and in my turn spoken fluently but for your deliberate denial'. Is it really plausible to say that all of these things that their child *could have done*, but for the parents' decision, are unimportant and the ability to do them and to experience them *counts for nothing*, such that its loss or absence is not a 'disability'? (Harris 2000: 97)

The implicit claim that Harris makes in this passage is that life without deafness is better than life with it. This is a relative claim because he is certainly not suggesting that life with deafness is not worth living. Nor is he making any judgements about those who are already deaf.

Activity:

Put yourself in the position of the fertility clinician or nurse counsellor dealing with Linda and Philip. How might you respond if you disagreed with the couple's decision. Would you:

- refuse to help them on conscientious grounds?
- respect their right to autonomy but express your disagreement with their decision?
- override their decision – refuse to implant the embryo with the gene for deafness, without informing the couple that other clinics might take a different view.
- respond in any other way?

Now, let's continue with Garanis-Papadatos' understanding of this case where she offers the opportunity to use our imagination and consider what Deaf existence might be like, an imagining that perhaps the clinician could not appreciate. We operate on the assumption here that imagination is not trivial when it comes to empathetically entering into another's moral or cultural perspective.

If you want to get deeper into this, some knowledge of the history of the Deaf is necessary. This community strongly argues that its members are not disabled persons but members of a linguistic minority with their own culture, history and language, a view also supported by many hearing people who live, work or are in any other way associated with the Deaf (Tucker 1998). A closer look at the attitude of the Deaf community towards the hearing world, reveals a conflict which has been alive for centuries now. It follows that the use of genetic technologies which can help a deaf couple to have a deaf child constitutes for them a way to help this community survive because it is their 'own' community: the desire of Linda and Philip to have a deaf child does not derive solely from their belief that the quality of their life is better than that of the hearing, but also from their feeling of belonging somewhere, of being members of a certain 'clan'. This kind of feeling is innate to a human being but it is a feeling that the hearing world has never been able to give to that part of the population which is deaf. So, it is not only the 'technical' aspects of hearing which influence the parents' decision making, i.e., that their child will be better off living in the Deaf Community and acquiring the knowledge of the sign language, but something beyond this – a fact which renders the doctor's position even more delicate. What he is dealing with is not solely personal, it is not an irrational or absurd wish but the expression of a deeper identity awareness and cultural feeling of a whole group of people.

The existence of PGD offers the potential parents and especially the woman a choice which is not only individual – to have or not to have a deaf child – but also social – with ramifications regarding the position of deaf people and all disabled people in society as a whole. The choice is related to the question of whether deafness is a disease that should be cured or a condition that should be respected as a cultural characteristic. The notion of choice sounds *per se* so beautiful and democratic. But,

- is this choice morally and legally trouble-free?
- is it really an autonomous one?
- does it actually enhance the democratic structure and the well-being of a society?
- is Linda actually 'choosing' to have a deaf child or is she succumbing to various pressures coming from her deaf environment?
- if a technical possibility is now available through PGD to determine certain features of your offspring, does this availability place pressure on special cultural groups to utilise it?
- let's think of this case with a variation: a deaf or a hearing woman trying with the use of PGD to avoid having a deaf child. Would you feel differently about that kind of choice? Consider why you might? Do you think there are similar problems? Why?

The issue of pressure being put on couples where PGD is available raises many thorny questions related to the right of the parents to make a choice, the nature of their responsibility and the extent of their power over any future children. Is it finally just a question of autonomy or is beneficence or the

well-being of any future child in conflict with parental autonomy here? (Garanis-Papadatos 2000a)

> **Activity:**
>
> In our earlier discussion around ideas about 'quality of life' we
> saw that it cannot solely be a judgment about the medical condi-
> tion of a person and that any such judgment encompasses a range
> of values that contribute to the complexity of the concept.
> Considering however, the dominant values in existing medical and
> social settings, do you think that the decision of Linda and Philip
> could be deemed irresponsible? What is your own view and how
> do you justify it?

Responsibility

So far, the concept of responsibility has loomed large in envis-
aging the future with reprogenetics. It focuses attention on
human beings as moral actors who think accountability
matters. Michael Yeo and Anne Moorhouse put it this way.

> Being morally accountable does not mean that one will always make
> what others, in particular those to whom one is accountable, believe
> is the 'right' decision. But it does mean being able to defend and
> justify whatever decision one makes...this helps us focus the task of
> ethics. Ethics is in the service of being accountable for our choices.
> Ethics at least helps us to make morally principled decisions, deci-
> sions we can justify with reference to the moral ideals and principles
> at stake and defend, if called upon, to give an accounting. So,
> although the real world pressures on decision-making are formidable,
> they are no excuses for failing to make morally defensible decisions.
> The person whom an accounting is owed is unlikely to be satisfied
> by the answer: 'I didn't have time to think about it'. (Yeo and Moor-
> house 1996: 31–32)

So a first view of responsibility sees it in terms of being
accountable for giving reasons, for acting according to some
principles, values or ideals that we can defend. One set of
such principles, respect for autonomy, nonmaleficence, beneficence and justice, offered by Beauchamp and Childress
(2001), has already been outlined in the Introduction. The
case of Linda and Philip illustrates the challenges of using
these four principles well. Re-read the case with them in
mind. Do they help to clarify the tension between the

fertility clinician and the couple? Linda and Philip thought that exercising their autonomy was also acting according to beneficence. Denying a child the opportunity to live in a Deaf Culture might have been interpreted by the couple as maleficence, causing harm!

Responsibility – more than one dimension

One way of being morally responsible in the broadest terms might be to utilise the four principles stated above in reflecting on dilemma situations. But responsibility is a concept with more than one dimension. On the one hand, it is partly defined in terms of certain roles or offices. Persons are often occupiers of particular social roles: teacher, nurse, parent, doctor, lawyer or combination of more than one of these. But, as we just saw in the previous section, our judgments about who is responsible for what on particular occasions always presuppose some set of background beliefs.

In the final chapter of this workbook we will consider PGD again, specifically in relation to the meaning and implications of 'social responsibility'. For now, however, we will examine the different kinds of responsibility that pertain specifically to health professionals and parents.

Responsibilities of health care professionals

Professional role responsibilities of a doctor would traditionally include:
- paternalism
- nurturing life
- minimising suffering
- correcting disease or malfunction which might interfere with human quality of life
- promoting patient's overall well-being and pursuit of human good.

One could see these responsibilities linked to traditional understanding of the goals and aims of medicine. But the question arises as to who interprets the meaning of 'well-being' and 'human good'. The professional responsibilities just cited here are undergoing steady change and are being interpreted anew as medical technology makes possible prolonged life even at the expense of patient well-being. Moreover, while the responsibilities of health care professionals link in with the understood goals of medicine, the latter too may alter as a result of several factors:

- sociocultural changes in patient expectations and values with respect to the doctor-patient relationship;
- development of medical therapies and technologies giving greater emphasis to preventive medicine;
- changes in professional status of nurses leading to encouragement of greater autonomy and resulting in a new distribution of role responsibilities
- litigation of doctors, nurses and specialists effecting possibly defensive attitudes towards patient rights.

It seems inevitable that these developments would effect changes in our understanding of professional responsibility for nurses and doctors.

In noting changes in professional responsibilities, there is a profound difference between the goals of curing infertility through a variety of therapeutic methods and trying to work with patients to ensure that embryos for implantation are (as much as can be determined) free of serious disease. Alternatively, as the case of Linda and Philip typifies, the goal of medical practice is dramatically changed when reprogenetic potential increases the probabilities of an embryo inheriting a specific feature or property usually thought of as pathological and undesirable. (For a further discussion of the goals of medicine, parental responsibility and the related conflict between the autonomy of parents on the one hand, and concerns about the welfare of the offspring on the other, see Chapter 7 on access to infertility treatment in this workbook.)

Maybe a fertility clinician would also see it as a 'goal of the fertility clinic' to ensure that all children who are born as a result of treatment in the clinic are without serious disease or pathological impairment. Here there may well be considerations to do with the clinic's reputation and commercial viability, both of which are likely to influence the clinician's perception of 'responsibility'.

Activity:

(a) Do you think the avoidance of serious disease is an appropriate concern for the clinician? Would it be a concern that might seriously constrain autonomy of choice of potential parents? Would it perhaps result in discounting some ethical perspectives of couples coming to the clinic?

(b) Keeping in mind Garanis-Papadatos' questions about the rights of parents to decide and the responsibilities of doctors, re-read the case of Linda and Philip, which shows how hard it is to say just where practitioner responsibility lies! It might be useful to read Alan Goldman's analysis of one of the features that we have already discussed in relation to the responsibilities of health care professionals.

In treating patients, doctors [and nurses] have extended their notion of harm from the merely physical to include potential psychological harm. Now they must extend it further into a fully moral notion according to which 'harm' consists in preventing an individual's pre-eminent values from being realised, or in violating a moral right. So this broader notion of 'harm' requires that a right to self-determination is recognised as fundamental. (Goldman 1980: 227)

(c) Take a few minutes to jot down how you see the responsibilities of a family doctor to a patient who seeks information on avoiding a heritable disorder. Is the doctor responsible only to her patients? To her profession? To society?

Responsibilities of Parents

One proposal from Jeffrey Blustein on parental duties and obligations claims that apart from minimalist duties to feed and clothe children, parents have at least two distinct sorts of duties towards their children.

- a duty to expose their children to the psychological conditions that facilitate development of the capacity for self-determination
- a duty to raise their children in such a way as to promote self-fulfillment: in this respect the primary goods of self-respect and health are essential. (Blustein 1982: 199)

Blustein would see that 'a full moral theory of parenthood presupposes some conception of the good life for mature human beings; for it is the purpose of childrearing to enable children to lead at least minimally decent and satisfying lives as adults' (Blustein 1982: 120). In this task, Blustein further argues that 'perhaps the most important parental duty is that of providing children with the kind of affectionate, appreciative, and supportive upbringing that gives them a sense of their own value and a confidence in their ability to fulfil their intentions' (1982: 129).

Parental responsibility on this view, challenges us to assess how this is shown by the use of NRTs. Some uses, drawing on PGD, seem to be circumscribing the value of the child in accord with parents' aspirations for an offspring with specific features. This concerns Onora O'Neill, too, who is definitely not sanguine about parenting in the new era of NRTs. Notice her links with the expanding terrain of choice over chance acceptance of parenting conditions. O'Neill's main concern is that 'as parental relationships are increasingly chosen relationships there may be a risk of sliding towards an increasingly conditional view of parenthood' (O'Neill 2000: 43). She argues that, insofar as children are products of complex genetic choices which attempt to exclude 'defective features' or 'disease states', acceptance or love may increasingly become conditional on the child having and displaying certain planned characteristics. Following on this, O'Neill raises doubts about the 'moral legitimacy' of PGD since it might adversely affect unconditional acceptance in parenting.

Activity:

(a) If Linda and Philip have a child who is not deaf, do you think that their acceptance would be different to this hearing child? Would their acceptance or love be conditional? How should the couple plan for the resulting child(ren) not knowing whether they will certainly have a deaf child or hearing?

(b) Do you think the 'good' parent has specific hopes and expectations for the features of their future children beyond the fundamentals of basic health? Is the self-determination choice of our couple in the case a 'responsible' choice as parents?

Give reasons why you think 'yes' or reasons why you would find difficulty in agreeing that their decision is 'responsible'.

(c) Are there aspects of the Linda and Philip narrative and the earlier case of Alice and Martin that you think have been overlooked here? What are these and why do you think they need to be addressed?

What of ethics if moral disagreement persists?

If you compare your judgments on the cases in this workbook with your colleagues and friends, it is very likely you will meet with disagreements. Disagreements are highly probable when a number of factors in moral issues make for moral complexity. An example of such complexity is certainly arguments about the relative benefits/deficits *and* the moral legitimacy of parents and health care researchers taking control or not taking control of the processes of genetic inheritance. Consider the following variation on the Linda and Philip case where parents make a decision for genetic transmission of a particular disease.

The Case of Mary and Robert

A couple knows that each of them is a carrier for the gene that causes Tay-Sachs disease. There is a 25 percent chance that any child they produce will have the disease. They ignore their doctor's advice not to conceive. They are deeply religious, and understand their religious beliefs to forbid interfering with the genetic endowment of the children they hope to conceive. They might explain their decision in words like these: 'We have thought about this for a long time. Given our religious beliefs, we believe it is wrong, in fact it is the worst sort of arrogance and impiety, to "play God" by trying to change the genetic endowment of our children. God has a plan for them, if in fact He gives us children, and we must not interfere with that plan. All human life is precious, and we must all die. We just do not know when we will die. (Andre, Fleck and Tomlinson 2000: 139)

NOTE: with Tay-Sachs, neurological deterioration begins at the age of approximately one year and continues until death by age five or earlier. The process is prolonged and grim with few compensations in the child's short life for the suffering it will endure.

Have you noticed that, in the three cases cited so far in this workbook that the understandings of 'responsibility' and disagreements that arose derived from background beliefs and deep personal convictions coming from various sources:

- convictions about a doctor's professional obligations to assist the birth of a 'healthy child'
- convictions about the imperative of 'biological' parenting
- convictions regarding 'quality of life' and of Deaf Culture
- convictions about the rights of parents to make choices about the 'outcome' of the child to be.
- convictions from religious traditions and institutions.

Moreover, in these cases, the stance taken by doctors or nurses involved encouragement sometimes, but also disagreement or disapproval. So the cases illustrate moral complexity and disagreement.

Activity:

(a) Do you think that the parents' appeal to the tenets of their faith is a persuasive defense for the moral justification of their choice? Do parents have a 'right' within their obligations to permit their child's suffering or death for the sake of protecting values central to their culture or religion or family?

(b) Can you think of ways you would negotiate the disagreements between health care practitioners and patients? Would you agree that, in the last analysis, parents have the right to choose and this right should be primary within provisions of democratic societies? Would you try to dissuade the parents' of their choices in any of the cases? Why?

Summary

This chapter has explained what is included in the concept of 'reprogenetics' and indicated some of the repercussions that advances in reproductive and genetic technologies have for our understanding and practise of responsible decision-making. In particular, we

- practised teasing out the values, ideals and principles that are 'at risk' in the case studies and considered the reasons that were offered for the decisions taken.
- clarified the meaning of 'quality of life' and discussed two of the components of responsibility: professional and parental.

– 2 –

Donor Insemination: Anonymity, Secrecy and the Right to Know

Objectives

At the end of this chapter you should have an understanding of

- the relevant similarities and differences between adoption and donor insemination (DI)
- the changing profile of donors
- the possible relationship between donor anonymity and secrecy in families and some concerns about the impact of such secrecy on all involved
- arguments for and against the disclosure or non-disclosure of information to children conceived through DI.

The first chapter of this workbook introduced the idea of responsibility in this age of genetics and new options for assisted procreation. We gave the term, 'reprogenetics' to this combined use of genetics and methods of assisted reproduction. Chapter 2, on donor insemination (DI), is also concerned with teasing out what is involved in responsible decision-making when a family chooses DI. Are parents obliged to tell children of their DI origin? What reasons can one give for or against telling children of their genetic parentage? Donor insemination is not a new reproductive technology; DI has been a practice for considerable number of years. However, there have been new developments in empirical research and philosophical argument about the issue of disclosing genetic origins to children born from DI. So, this old practice is revisited here with new data and viewpoints for you to consider and question.

Before we begin exploring questions of responsibility in donor insemination, take a few minutes to read the following case study about a family born through DI and answer the questions below.

The Case of Sarah and John

When Sarah and John, a married couple, discovered that John was **azoospermic**, Sarah wanted to try for adoption but John preferred the idea of donor **insemination** treatment since he *very much wanted to cover up and make us look just like a normal family'*. Today, they have two sons, one aged 3 _ and one aged 7 months, both conceived by donor insemination (DI).

The couple disagrees on how open they should be about the method of conception. Sarah has told her immediate family, but none of her friends know. This makes her feel uncomfortable, particularly when they comment on her sons' resemblance to her husband. *'I often think, how can people say he looks like John when he's not even his genetic child? It's such a difficult thing to contain myself from saying anything and it's going to carry on like that for the rest of our lives.'* However, she feels that to say anything would be to *'dishonour my husband'*.

John's parents have not been told anything about the donor insemination since he believes that it would *'destroy them to find out that their own son was unable to father his own genetic children'* and that they might, as a result, turn against their grandchildren. His opinion is that they should keep it to themselves since *'it's all very well telling everyone but the kids might not want everyone to know when they get older'*.

The biggest point of contention is whether to tell the children. John feels it would not be beneficial to tell them since the donor is untraceable: *'I can understand if it was adoption because then they can go and find their parents if necessary, find their background but with this they've got no way of finding anything'*. He is against donors becoming identifiable since the family is reasonably wealthy and he worries that the donor might try to claim some money from his sons if they contacted him. Another big concern is that once the boys know about the treatment, they might tell his parents. Sarah argues that her priority is to protect her children not John's parents. She is adamant that she will tell her sons the facts about their conception in order to be honest and *'share the joy of what has turned out to be an absolutely fantastic thing for me'* even if it means going against her husband's wishes.

The problem is with knowing how and when to tell the children: *'I don't want it to be an issue, I want it to be a natural piece of information'*. Unlike her husband, she plans to lobby for the right of donor offspring to be able to identify their genetic parents stating that *'my child has a right to know his physical, genetic origins … just a simple basic human right'*. Without the information to forge that *'missing link'*, she feels that it is *'always going to be more complicated than just a straightforward natural family'*. (MacCallum 2000a)

Notice that the case makes no mention of whether the couple had any offer of counselling prior to receiving DI. Practicing psychologists might argue that counselling would have helped prevent family tensions which are revealed in the case. Some would argue that the couple's tension and anxiety over disagreement on anonymity is needless and unnecessary. With good counselling prior to the decision for DI, perhaps this difference of viewpoint could be worked through with good benefit for the resulting child as well.

Activity:

How would you view this means of adjudicating the couple's differences?

- For persons using DI, should counselling be made a prerequisite for donor insemination?
- If a couple talks to their family doctor about considering the use of DI, do you think the doctor should recommend that the couple seek counselling to clarify their decision? Or should the doctor take this on as a responsibility?

> ● When the couple approach the sperm bank or fertility centre for sperm donation, would you expect an opportunity of pre-insemination counselling be offered to the couple to try and off-set some of the family problems that might arise where a couple differs about the desirability of telling the future child.

Now read Fiona MacCallum's commentary on this case and see whether the empirical research now available about DI families influences you in developing your own views on this case.

Commentary on the Case of Sarah and John
Fiona MacCallum

Until the last few years, couples who conceived using donor insemination were not encouraged to be open about the method of conception and studies show that the vast majority did not tell their children (Cook et al. 1995). Legislation such as the United Kingdom Human Fertilisation and Embryology Act in 1990 enabled parents to register the mother's husband as the father regardless of genetic links, thus legitimising the relationship between the child and the parenting couple and protecting donors from claims to paternity. Before this, the issue of legal parentage may have contributed to the secrecy of DI, i.e., mothers may have registered their husbands as the father anyway and felt that this meant they could tell no one about the DI. This attitude has been changing recently with growing social awareness of the use of donated gametes and the establishment of support groups which encourage telling. (One such support group is the DC (Donor Conception) Network in the United Kingdom, which was set up by a couple who had conceived using DI and wished to tell their children. Similar groups exist in the United States, Canada and Australia. These are generally set up by DI parents or DI offspring rather than clinics.)

Problems can arise when some of the family members are not told, in this case, John's parents. John is assuming that they would be devastated. However, they have not had a chance to prove him wrong. For many men, infertility still carries a stigma and it may be that John is projecting his own feelings of inadequacy onto his parents. Donor insemination does not cure male infertility; it simply makes it possible for an infertile man to be the nurturing father to a child. He must still come to terms with the fact that he will never have his own genetic child. (MacCallum 2000b)

Activity:

(a) In your own experience of your society's practice of DI, do you find that this attitude of not telling DI origin is changing?

(b) Does infertility carry a stigma only for men? Is it considered a stigma for women too?

(c) Following on your response to (b) above, to what extent is infertility given a different meaning, attributed with different seriousness (for men or women) according to the society or culture in which it occurs?

(d) How would you address John's concerns and fears with regard to his parents finding out about his infertility?

Now return to MacCallum's commentary.

Sarah does not feel comfortable being dishonest with her children. Evidence from family therapy shows that secrets can be harmful, in that they set up boundaries between those in the family who know and those who do not (Papp 1993). Children can become aware that there is something hidden, e.g., when Sarah hears comparisons between her sons and her husband, they may pick up her discomfort and become confused or anxious. Since Sarah's family is aware of the DI, keeping it from her sons would leave open the possibility of later inadvertent disclosure that could disrupt the bonds of trust between the children and their parents. In adopted families, children fare better when parents communicate openly about the adoption so it may be that the same is true for donor conception children (Brodzinsky et al. 1998). As yet, there is no empirical data on the effects of disclosure to children, so we cannot automatically assume that it will always be beneficial. Anecdotal evidence from adult DI offspring who have found out about the DI in later life suggests that non-disclosure has had a negative impact on their well-being but these adults are not necessarily representative of all donor offspring (Turner and Coyle 2000). Studies of young children born by DI who have not been told show no evidence of any harm caused by keeping the method of conception private although this may change as the children get older (Golombok et al.1996).

Activity:

(a) Do you think that secrets are harmful in a family? How important do you consider the fact that one part of the immediate family knows and the other does not? How important is the fact that there is disagreement between the parents?

(b) Think about the analogy with adoption. What similarities and what differences do you see?

The following are some suggestions about similarities and differences between adoption and DI birth.

- In both cases there is no complete genetic link between parents and offspring, in DI, however, there is a link with the mother.
- In both cases the social parents want and care for the child. In adoption, however, a child could discover that there may have been some kind of rejection by the genetic parents.
- In both cases there are more people than the two parents who know about the adoption or the DI. It is much easier, however, to keep the secret of DI since people have seen the mother pregnant and if the two parents agree, the child can always appear as their own.
- In adoption, the files kept by the adoption agency can lead to the genetic parents, while in DI, even though there are files about the medical history, etc., of the donor, his personal identity is not always known.
- In adoption there is a legal procedure involved, and children might come across the legal documentation at some point in their lives. In DI there are files but it is most unlikely that anyone will come across them, especially if they are not looking for them.

Now consider MacCallum's concerns about if and how to give children information about their DI conception.

The continued anonymity of donors seems to support secrecy in families. The right to know one's genetic origins has been highlighted in the adoption literature, and it is hypothesised that not having this information may be detrimental to emotional and psychological development. However, the counter argument could be made that secrecy and anonymity are in the interests of the child herself, since they ensure her well-being (Council of Europe 1989).

Many parents, like John, cite the unavailability of genetic information as a reason not to tell the child. Sarah feels that the children have a right to know their donor, but this must be weighed up against the donor's right to privacy. One study of sperm donors found that two-thirds of them would not donate if there were a chance of their offspring tracing them (Cook and Golombok 1995). However, a new style of recruitment policy could be put in place to attract a different type of donor who would be willing to be identified. This has been done with some success in Sweden (Daniels and Lalos 1995).

If the parents do decide to give the children information about their conception, they come up against the issue of when and how to tell. This can be difficult due to a lack of established scripts although there are now some books available that tell the story of DI in a way which children can understand (Cooke 1991, Schnitter 1995). It appears it may be preferable to start giving information when the children are at a young age in order that it can be processed in a matter of fact way rather than coming as a shock to the child (Rumball and Adair 1999).

Activity:

(a) If you were helping someone to tell their child that they were conceived through DI what narratives would you suggest?

(b) Who do you think is benefiting from the maintenance of secrecy?

The genetic or the nurturing father? John's parents? The child? The family as a whole?

Whose interests do you think should prevail?

How do you defend the position that you take?

Having introduced the case of the family born of donor insemination, we would like you to read a paper on DI by Heather Widdows. Compare MacCallum's commentary and your own view with the position expressed by Widdows.

Secrecy in Donor Insemination
Heather Widdows

Secrecy has been an integral part of donor insemination since its beginning (reputedly in 1884) (Daniels and Haimes 1998, Gottlieb et al. 2000). Recently, attention has been given to the possible adverse effects of secrecy and, accordingly, the practice of secrecy in DI has been questioned. This paper analyses the reasons which have been given for and against secrecy and will consider the effect that changing the practices of secrecy might have on DI.

Many explanations have been put forward for continuing the practice of secrecy in DI. These range from patient confidentiality to social reasons such as the stigma attached to illegitimacy (Pfeffer 1993). Secrecy has become, either through time or design, not simply an addendum to DI but part of the structure of the procedure. The integral part which secrecy has played in DI makes exploration and analysis of this topic difficult. Not only is DI less in the public eye, and so less discussed than other Artificial Reproductive Technologies (ARTs) such as In Vitro Fertilisation, but also secrecy 'covers its own tracks', in that, little evidence exists regarding the effects of secrecy on families who have used DI to conceive: parents are unwilling to talk about their use of the procedure, and offspring of DI are unable to as they do not know the manner of their conception.

I explore the issues of secrecy, focusing on two areas: anonymity of donor and non-disclosure to the child. I will give the arguments both for and against secrecy and assess them.

Donor anonymity

The ethics of donor anonymity in DI has become prominent over the last two decades, and has been brought into relief by the removal of donor anonymity in certain countries. Changes to DI practice in these countries provide a context in which some of the usual justifications for secrecy can be assessed. In addition, recent advances in genetics have strengthened claims that knowledge of one's genetic parentage is an important part of understanding one's own identity (at least medically). Moreover, such advances have also reduced the likelihood of keeping non-genetic parentage secret. Taken together these factors have led many to re-examine the traditional assumption that donor anonymity is the 'self evident principle of DI' (Bateman 1998: 119).

Activity:

What is your opinion about the statement that 'genetic parentage is an important part of understanding one's own identity'? To what extent is genetic information necessary? Is it knowledge of genetic parentage or simply one's genetic make up that is important in understanding one's medical identity?

First, the supposition that if donor anonymity were removed then donors would no longer be willing to donate sperm can now be tested against the evidence which is emerging in countries where anonymity has been removed. The most detailed evidence comes from Sweden which was the first country to move away from donor anonymity in 1985 (Swedish Law of Artificial Insemination, March 18, 1985, no 1140/1984). The Swedish legislation allows DI 'children' access to identifying information about donors when they reach maturity. Many predicted that outlawing anonymity and making the donor identifiable would result in a dearth of donors and even the end of DI. However, such predictions proved alarmist. After the introduction of the law the number of donors did initially decrease, as did the number of DI births, and simultaneously the number of couples travelling to other European countries for DI increased (Daniels and Lalos 1995). At first sight such evidence appears to suggest that both donors and potential parents were uncomfortable with the removal of donor anonymity: donors were less willing to donate and parents were choosing to go to countries which continued the practice of donor anonymity. However, this is not the only explanation and it is arguable that other factors were at work.

For example, Daniels and Lalos (1995) suggest that one alternative explanation for the decline in donors derives from changes in legislation in Sweden regarding the compulsory testing of semen for HIV before and after six months of cryopreservation. These changes resulted in private clinics ceasing to offer DI, which meant that donors were no longer required, and that couples had no choice other than to seek treatment elsewhere. To support the claim that removal of donor anonymity does not inhibit donation, Daniels and Lalos surveyed the numbers of donors in the eight DI programmes in Sweden between 1989 (the first year that figures were recorded) and 1993. These figures show a steady increase in the number of donors, and an overall increase of 65 percent. Unfortunately, there are no comprehensive figures from before the 1985 legislation. However, Daniels and Lalos conclude that 'despite the limitation, it is clear that the number of available donors is increasing' (1995: 1873). To support this conclusion, they cite statistics from the University Hospital of Northern Sweden, which had collected donor figures both before and after the introduction of the law. These figures show that the number of donors pre- and post-legislation remained static, and later (coinciding with high profile recruitment campaigns) the number of donors began to steadily increase, thus supporting their claim that despite the removal of anonymity donor numbers are increasing.

From the evidence above it can be concluded that removing donor anonymity would not stop donors coming forward, but it would cause changes to the structure and current practice of DI. The two most notable changes in Sweden were changes in public perceptions of DI, and, more crucially, changes in the type of men coming forward to donate sperm. In order to encourage donors to come forward, new strategies were needed and high profile recruitment campaigns were introduced. These campaigns raised public awareness of DI and thereby reduced the secrecy surrounding the procedure. The more fundamental change was to the type of donors willing to donate. Before the removal of anonymity, donors tended to be students who were motivated primarily by money, whereas donors recruited after the change in legislation tended to be older, married men, who were motivated altruistically by a desire to assist infertile couples (changes also reflected in studies in New Zealand and Australia) (Daniels and Lalos 1995). Thus, although the predictions that removing anonymity would stop sperm donation (and so DI) have proved false, notable changes have occurred. In one sense the predictions were correct, in that the donors who donated before the passing of the law (the anonymous donors which the predictors had access to) did cease to donate once anonymity was removed. However, this proved to be unimportant in terms of the overall number of donors, as other donors were prepared to become non-anonymous donors.

Activity:

(a) Do you think that the Swedish experience would be the case in your country? Would there be a shift in the donors' profile if donor anonymity were removed? Do you think that older men with families would be willing to donate semen, undertaking the risk of the disclosure of their identity at some point of their life?

(b) If donor anonymity remains in place in your country or elsewhere, do you think it makes an important difference whether the sperm donor is altruistically motivated or not? If you think it does make a difference, can you explain what that difference is? In other words, is altruism as a feature of donors only important if anonymity is removed?

(c) Do you think that a society can defend its decision not to give its children the option to know their genetic father's identity?

Now return to Widdow's paper and to a summary of her position so far.

In sum then, the first reason for continuing anonymity is not substantiated; donors will not stop coming forward. Hence only the second reason for insisting on anonymity remains, namely, that anonymity ensures that donors have the 'correct' attitude to the procedure. Before the recent questioning of anonymity, the secrecy involved in the process of DI was taken for granted and was unquestioningly assumed to benefit all concerned (donor, parents, child and doctor). In such a framework, it was in the interest of all parties to keep their involvement secret, and anonymity was seen as safeguarding secrecy for both the donor and the parents. The correct attitude of the donor was thought to be detachment; the donor should not wish to know anything about, or have any contact with, his potential progeny (Pennings 1997). Anonymity guarantees that the donor provides his semen – the raw material of DI – and that this is the end of his involvement; there is no hope of any future knowledge of, or contact with, any offspring resulting from his donation. This attitude is further enforced by paying the donor's 'expenses' (importantly, at least in the United Kingdom, expenses not payment). Such reimbursement provides some reciprocity which, at least symbolically, implies an end to the encounter. In addition to providing a symbolic reciprocal act, the money which changes hands does provide motivation for some donors. Although the level of expenses is intended to be below the level of inducement, for many young men (characteristically students) the expenses are sufficient to function as inducement to donate (Daniels and Lalos 1995, Pennings 1997a). Indeed, it could be argued that this perception is the one intended, as paying expenses encourages the sense of conducting a transaction, which lowers any possibility of the donor feeling any entitlement to future information or contact with any possible children. This is the traditional model in which the donor simply provides:

> genetic material in order to enable others to fulfil their wish to have a child. The donor is an outsider who has no rights or responsibilities in the newly created family. The procedure ... completely severs the link between the donor and his genetic material and thus, indirectly, isolates the donor from the recipients. (Pennings 1997a: 1842)

Activity:

(a) Is this the model of sperm donation in your country as well? Is there a payment to the donor, or only reimbursement of expenses? What is the motivation of donors?

(b) Is there a model of egg donation in your country? Is there a payment to the donors and what motivates them?

There are four usual models of egg donation:

1. women who undergo IVF treatment donate eggs which will be used by other women in the same situation. The donors have a financial gain from this procedure since there is a reduction in their hospital expenses. (This process is sometimes called egg-sharing.)

2. donors are recruited through advertisements and they are usually paid for their services

3. a woman who is going to undergo IVF treatment brings her own donor with her.

4. a woman who is going to undergo IVF treatment brings a donor who is exchanged with the donor brought by another woman so that anonymity is respected.

List the similarities and differences that you see between the models of sperm donation and egg donation.

Now read Widdow's claim that the framework of sperm donation in many countries is currently undergoing a radical transformation.

Removing anonymity fundamentally alters the framework in which DI has been practiced and threatens the long established culture of secrecy. Instead of attracting donors who wish to have no contact with the offspring their sperm are used to create, donors are attracted who do not feel that anonymity is important and therefore are willing for their donor-offspring to know who they are and perhaps even to be contacted by them.

The conclusion which must be drawn is that those who support the continuing practice of donor anonymity do not fear that there would be no men willing to donate, but rather that these donors would be the 'wrong' type of donor. Changing the type of donor – from anonymous and financially motivated to identifiable and altruistically motivated – threatens the present model of DI. While DI could still be used to solve childlessness, the ethos of the procedure would be very different. A prime concern would no longer be to keep the procedure secret and to keep the donor separate from the couple. In particular, instead of enforcing the pretence of a 'normal' family – by which is meant the traditional (and many would argue outdated) model of father and mother and genetically related children – the change makes openness possible. Indeed, changing to identifiable donors implies disclosure to the child. For while one can inform the child of his/her donor conception, if donors are anonymous (the child would simply know s/he was conceived by an anonymous donor) one cannot give the child identifying information about the donor unless the child has first been informed of his/her status as a DI child. Thus, removing anonymity challenges the culture of secrecy. While anonymity is in place, parents may feel that there is little point in revealing the fact of DI conception to the child as no information about the genetic father is available. However, on the removal of anonymity the reverse is the case; there is no point removing anonymity unless parents tell their children. Hence, removing anonymity can be seen as putting pressure on parents to reveal the mode of conception to their children.

Secrecy, then, in the form of donor anonymity, does not protect donors as a homogenous class, but only a certain type of donor and thereby a certain structure of DI. Removing anonymity affects the culture of secrecy which has been at the heart of DI, and implies significant changes in the way DI is regarded by users and by society as a whole. Not only does removing anonymity put pressure on parents, but it also presumes that society will accept DI as an alternative means of family creation in a similar manner to the way that other ART's and adoption have been accepted. It could even be argued that removing anonymity introduces the presumption that there should be a relationship between donor and donor offspring, something that is anathema to the traditional concept of DI.

Secrecy and the family

The second key issue is secrecy in the family, more specifically, the non-disclosure by the parents to the child. The practice of secrecy has been defended on the basis that it protects the family: its individuals, their relationships with each other and the family unit as a whole. An important reason which is given in defence of secrecy in the family is that it protects the family from the stigma of male infertility (Klock et al. 1994, Nachtigall et al. 1997, Lasker 1998). Fear of admitting male infertility is cited as a key reason for non-disclosure and this seems to be supported by the evidence, in that those who use DI to overcome male infertility are less likely to disclose to the child. For example, couples who use DI because of vasectomy, or to avoid passing on a genetic disorder, are more likely to disclose than couples in which the man is infertile (Nachtigall et al. 1997, Lasker 1998).

Crucially, when couples use DI, unlike all other ART's, there is no doubt that it is the man, and not the woman, who is infertile. For example, even though IVF is often used for men with low sperm counts (either naturally or after vasectomy), the focus and presumption of infertility rests with the women (Spallone 1989). DI reveals male infertility, and so the 'cultural assumption of infertility being primarily a female problem is violated for these couples' (Lasker 1998: 14). Male infertility does go some way to explain why couples do not disclose to the child, and why there is less open acceptance of DI at a wider society level. This is linked to the wider topic of the importance of heredity and genetic relatedness.

Activity:

Some commentators claim that children feel the cloud of secrecy in the family. But, consider whether you agree with this? Do you think the child is affected adversely by secrecy over DI?

Historically, the claim that secrecy is in the best interest of the child was a strong argument because secrecy protected the child from the stigma of illegitimacy. But illegitimacy is no longer a major concern, at least in most Western countries, especially when it is due to the use of ARTs. Many countries have begun to remove the language of illegitimacy from their statutes. If this is so, then the claim that secrecy is in the best interests of the child must be defended with other reasons. The reasons that are usually given to defend secrecy are:

1. that not knowing about the DI conception guarantees the child stable and 'normal' family relationships, and prevents any uncertainty about identity (which could result from knowing about the DI conception);
2. that openness is damaging to the child's relationship with his/her parents, especially with his/her social father (even to the point of rejection of the social father in extreme cases).

Alternatively, the arguments for openness are, firstly, that knowledge of one's genetic origins contributes in a significant way to identity formation, and secondly, that secrecy is damaging for the family as a whole. I will present each of these arguments in turn.

1. The importance of 'roots'

First, the suggestion that keeping the mode of conception secret has a positive effect on the child by preventing any questioning about identity has recently been heavily criticised. Critics argue that knowing one's biological and genetic heritage is of fundamental importance to identity, and indeed such is the presumption behind the change in the Swedish law, and the more open practices of other countries such as Australia and New Zealand. This perception is echoed at the lay level and there is a general agreement that 'roots', in some form, are held to be important (Edwards 1998)

To support the hypothesis that knowing one's genetic heritage is important, an analogy has often been drawn with adoption. The ethos of adoption has changed dramatically over the last fifty years, from one of secrecy to one of openness. Those who use this analogy argue that the same thinking can be applied to the 'right' of a child to know his/her genetic parents in DI. However, although there are obvious similarities between adoption and DI – viz. at least one of the child's social parents is not the genetic parent – the analogy with adoption is frail. This is for the following reasons: (a) the DI child is biologically linked to one parent (both genetically and gestationally); and (b) the DI baby has not been 'given away' and therefore does not have a history of rejection to resolve (as might be the case with adoption). Thus, although there are similarities it would be wrong to regard this analogy as clinching the argument for openness in DI. Nonetheless, there are arguments for openness which are used in adoption which do have significance for the case for openness and thus merit exploration.

The most obvious parallel concerns identity: a 'right' to know one's roots, for both emotional reasons (such as discovering the kind of person one's 'father' is and knowing the reasons why he chose to donate) and for practical reasons (such as medical, in particular genetic, reasons). This argument concerning 'roots' and identity has considerable emotional pull, and whether one accepts it or not largely depends on one's view about the importance of genetic relatedness. In addition, there are many cases where genetic identity cannot be known, making arguments for openness, which are based simply on this premise, tenuous. One could claim the knowledge of genetic parentage is desirable, but to claim it is an essential component in forming a stable identity is an exaggeration. This said, there is no doubt that keeping genetic history secret will become more difficult as the genetic revolution continues. The very nature of genetic testing is such that it yields information about genetic relatives, so, by mere force of circumstance, genetic relatedness (or at least non-relatedness), and hence identity, will be revealed. Consequently, and for purely practical reasons, maintaining secrecy in DI may prove impossible. Such a scenario would force openness and thus a re-evaluation of the significance of genetic relatedness and what is meant by 'family relationships'. In sum, the argument for revealing genetic knowledge is important for identity is not conclusive, although it may gain strength as knowledge of genetic heritage becomes more important for health.

Activity:

(a) You have given a lot of thought to the importance of heredity and genetic relatedness through your reading. Before you go on with Widdows' paper, could you now answer whether it would be in the child's best interest to know the genetic father? Think of it from the medical point of view, as well as the psychological and social view.

(b) How far can a 'right' to know one's roots be established? Think about those who do not know their genetic heritage due to reasons other than adoption or DI such as war, rape, etc. Do you believe they have an identity problem or is it an exaggeration that knowledge of genetic parentage is an essential component in forming a stable identity?

2. The damage that secrecy can do

The second argument for openness, which applies as much in DI as it does in adoption, is that secrecy is damaging for the family as a whole. The traditional DI assumption is that secrecy protects the family unit by ensuring that the family seems 'normal' to family, friends and society and appears the same as genetically-related families. The counter-claim, that secrecy is damaging to the family, which is used to support openness in adoption, can be applied to DI. If it proves to be the case that secrecy is damaging to the family and so to the best interests of the child, a crucial justification from maintaining secrecy will be undermined.

Two main reasons are suggested as to why secrecy is damaging to the family unit: (a) that the secret will unintentionally be revealed, and (b) that keeping secrets within a family is harmful in itself. The first and most obvious reason is the danger that the secret will come out, either directly, when it is told, or indirectly, in that the child growing up will form certain suspicions (Snowden and Snowden 1998). Most couples who have used DI to conceive have kept it secret from their offspring, yet they have tended to tell at least one other person. Given that these people are likely to have told one further person, it is probable that far more people know than couples are aware of, all of whom could potentially reveal the secret (Nachtigall et al. 1997). Consequently, the secret is far more likely to 'get out' than the parents imagine. If this happens the chances of a breakdown in the relationship between the parents and the child, even to the extreme point where the child rejects the non-genetic father, are much greater. This reason for rejecting secrecy is relatively uncontroversial as all accept that an accidental revelation of DI conception is clearly not in the best interests of the child. All couples accept this danger, and accordingly weigh the risk of exposure against the benefits of continued secrecy.

Activity:

(a) The traditional DI assumption is that secrecy protects the family unit because it seems 'normal' to relatives, friends and society. Consider other reasons too. Does the family feel 'normal' by ignoring the distant fact of DI? Or is this fact vivid in their every day life no matter how much they try to ignore it?

(b) Compare the danger of revelation of the 'secret of adoption' and the 'secret of DI' as well as the risk the couples undertake in both circumstances. You might consider the following dangers: i) revelation because of family talks and ii) revelation because of talks at school. Are these two dangers real for both adoption and DI? Isn't it true that in adoption the danger of revelation at school is greater because more people know about it? So, do you think that the risk the parents undertake in both circumstances is the same?

The second reason for rejecting secrecy is more contentious, namely, that secrecy is damaging in itself; that the simple awareness of a secret, even if it is never exposed, is harmful. Proving such a claim is difficult, not only because there is no evidence one way or the other, but also because such an absolutist position is so controversial. One possible way of approaching this issue is to consider the somewhat approximate case that lying, rather than simple non-revelation, is harmful. Making this adjustment is open to criticism, as most contributors in the field would argue that non-revelation does not equate with lying. However, in the case of DI it is possible to argue that keeping the mode of conception secret would probably necessitate lying, and even repeated lying: during the procedure (regarding time taken from work), at birth (regarding the identity of the father), in response to childhood enquiries (to the child himself), and so on. Moreover, it seems fair to conclude, that the need for lying increases as public awareness of ARTs grow. While it may have been possible in the 1940s simply to lie during the procedure and when registering the birth, it is far less likely in the present climate that one will be able to avoid lying to the child. Children are increasingly likely to ask questions such as, 'Mummy, was I born like that? How was I conceived?' Given this, and for the purposes of exploring the issue, lying rather than non-disclosure and its effect on the family and the child will now be discussed.

One of the best known formulations of the position that lying is wrong is a type of argument found in the eighteenth century philosophy of Immanuel Kant (Kant 1994 [1799]). For Kant, lying is never morally justifiable; it is wrong in all circumstances. No consequences or circumstances would justify it; it cannot be justified even for the most altruistic of reasons. The reason Kant gives for this position is that lying threatens the autonomy of moral agents by reducing their capacity to make rational and autonomous decisions and this capacity is fundamental to being human. Keeping the truth from a person creates a power imbalance which results in the 'lied-to' not achieving her full status as a moral agent, as a possessor of freedom and reason.

In order to justify lying, one has to adopt a non-Kantian position where one can claim that lies are justified if the consequences are beneficial. Moreover, if the lie is in the best interests of those lied to, a lie is not only justified, but even a 'good'. From such a standpoint, 'paternalistic lies' are justified on two grounds:

1) first, to protect the interests of those doing the lying;
2) second, because they are in the best interests of those lied to.

If these criteria are fulfilled then it is assumed that implied consent is given by the person being lied to.

With regard to DI then, those who advocate secrecy claim they are justified by the good consequences – a 'normal' family in which all the relations are kept intact – and because lying is in the best interest of the child, who it is assumed, would prefer not to know. So, implied consent from the child is assumed to be given.

Yet even if one were to adopt this position, and say that lying has no negative value, it is still not clear that the consequences in the case of DI actually do justify secrecy. To claim that lying protects the child of DI or, is in the best interests of the child, is doubtful. The situation is complex. It remains open to debate exactly what is in the child's best interests. As yet there has been no case in law where best interests include knowing one's genetic parentage.

Activity:

(a) Consider the concept of the 'normal family' used in the paper. Do you think the concept of 'normal family' has changed in recent years? If the concept has changed, what factors do you think are causing such change?

Take some time to reflect on the different forms that family life takes in different countries, classes, religious, ethnic, social and age groups. How many 'normal' families do you know?

(b) Before you read the conclusion of Widdows' paper, do you feel you have reached a conclusion yourself? Could secrecy about DI in a family be justified in philosophical and practical terms? If yes, how?

Conclusion

The arguments regarding donor anonymity are ultimately concerned with the type of procedure that DI should be. Those who wish to maintain anonymity wish for the culture of secrecy to continue and for donors to remain completely separate from the parents and the resulting child. Those who argue for removing anonymity are really hoping that this will result in a change to the wider practice, and that parents will reveal the fact of DI conception to their children. In practice this has not happened, and parents are still unwilling to disclose to the child even in countries which have removed donor anonymity (Gottlieb et al. 2000). Without parental willingness to disclose to the child, making the donor identifiable is largely immaterial. Drawing firm conclusions about whether or not non-disclosure to the child is in the child's best interests is difficult. This is largely because there is very little empirical data either way. Given this, it may be that the best, though unsatisfactory, solution is to leave decision-making entirely to the parents' discretion.

However, if decisions about disclosure are going to be left up to parents it could be argued that all parents should be allowed to choose an identifiable donor. Without this choice they may feel that although they can disclose the child's DI status, if the donor cannot be identified, the child would be frustrated at the lack of further information. Conversely, parents who wish for total secrecy may wish to choose an anonymous donor. Arguably, if the choice really is to be up to the parents, this option should be available.

However, this said, changes in the law regarding anonymity are moving away from one formulation of the practice to the other. By making such changes in legislation the lawmakers are suggesting that one course of practice is preferable. In this case, the preferred option suggested is that openness is better than secrecy. In fact, what has informed the changes in the laws of certain countries is the belief that it is important for donor offspring to have access to information concerning their genetic heritage. Moves to remove anonymity replace the presumption of secrecy with one of openness. Such a change is a major shift in the underlying ideology of DI and in the ethos surrounding the procedure. The effects of this change can already be seen in the more positive attitude of the public to DI as shown by the number of willing identifiable donors in countries which have abolished anonymity. This may eventually even bring about a change in parents' willingness to disclose to the child.

Finally such changes in policy and all of the above arguments may simply be surpassed by increases in genetic testing and knowledge. Such practical changes, which reveal genetic non-parentage, will force openness, whether or not parents wish it and whether or not psychologists, doctors and philosophers think it is beneficial.

Discussion

There are four main points in Widdows' conclusion:
1. Without parental willingness to disclose to the child, making the donor identifiable is largely immaterial.
2. If decisions about disclosure are going to be left up to parents it could be argued that all parents should be allowed to choose an identifiable donor. Conversely, they should be allowed to choose an anonymous donor.
3. Moves to remove anonymity serve to change the presumption of secrecy into one of openness.
4. Changes in policy and all of the above arguments may simply be surpassed by increases in genetic testing and knowledge.

The question at issue here is whether or not the responsibility regarding information about their children's conception should be left to the parents. Shenfield and Steele (1997) say that, in the face of lack of evidence of the consequences of secrecy or openness, future parents are best placed to decide on this matter for their potential children. Pennings (1997b) also proposes the 'double track' policy, i.e., let the parties decide for themselves. Donors may choose between anonymity or identification and recipients can opt for an anonymous or identifiable donor.

If the responsibility about secrecy or openness in DI is going to be left to parents, we would argue that any state has to form a system where:

(a) donors can decide whether they want to be anonymous or not,

(b) all their medical data be kept one way or the other,

(c) counselling is provided in all cases and,

(d) the civil law is changed accordingly.

However, even granting these conditions, the following quotations from people who were conceived through DI should give us pause:

> When secrets are kept, the children often grow up sensing that something is different within their family. The funny thing is that this is not necessarily due to what their parents do say, but as a result of what the parents don't say. For example, they never say, 'You've got your father's eyes and your grandmother's personality" Lauren Taylor, aged 23. (Donor Conception Group of Australia 2000)

> I long to know who my biological father is, and to meet and speak with him at least once. I search for my half-siblings in other people's faces. I want to know the missing part of my family history, but more than anything I need to know the other half of my ethnic background. Now that some of us are adults, and in fact older adults, it is time for our voices to matter. It's time to take a serious look at the true realities and implications of creating non-biological families with secrecy. We have a right to know our identity and to grow up in truth.' Lynne Spencer, aged 42. (Donor Conception Support Groups 2000)

It might well be argued by the people cited above that if we leave the choice of anonymity to donors and the responsibility of openness or disclosure to parents, we may also be leaving many of those who have the most at stake with very little decision-making power at all.

Summary

This chapter explored the ethical issues that arise in the context of donor insemination. Specifically, we investigated the arguments for and against maintaining the anonymity of donors and secrecy within the family. Below we list reasons given for and against secrecy. Can you give reasons other than those mentioned below?

Arguments for and against Secrecy

For

✔ not knowing guarantees the child stable and 'normal' family relationships, and prevents any uncertainty about identity (which could result from knowing about the DI conception)

✔ openness is damaging to one's relationship with one's parents, especially with one's social father

✔ ?

Against

✘ knowledge of one's genetic origins contributes, in a significant way, to identity formation;

✘ secrecy is damaging for the family as a whole since it involves them in ongoing collusion to keep genetic origins unknown.

✘ ?

Activity:

(a) Have you thought of other arguments to defend secrecy or openness about DI origin? We said at the outset of this chapter that DI was an established method of assisted procreation but that new arguments are being offered to challenge the assumption of secrecy within families about DI origin. Do you think the arguments have been convincing?

(b) The task of evaluating the pros and cons of any position is an important skill to be learned from the activities of the text. There is also the further difficult task encouraged in this text, that of coming to your own considered position after weighing the pros and cons. After reading the arguments in this chapter on DI, do you think you could formulate your own position?

– 3 –

Surrogate Pregnancy

Objectives

At the end of this chapter you should have an understanding of

- the motivation of the surrogate mother
- the significance, legally ethically and socially, of the quality of relationship between the surrogate mother and the commissioning couple
- the significance of biological and social ties in relation to determining parental status and responsibilities and our understanding of 'family'
- the impact of the surrogacy arrangements on the welfare and well-being of children and surrogate mothers

In Chapter 2 we looked at the practice of donor insemination and the ethical issues concerning donor anonymity and secrecy in the family that are associated with it. This chapter is concerned with the issue of surrogacy that continues a discussion of the option of using a 'donor component' in reproduction. A woman is a donor in a novel sense; she donates the use of herself, her womb and her labour to provide a child for another couple who might otherwise remain childless. She may or may not get money for this transaction. Surrogacy is, primarily, a social solution (an arrangement between collaborating individuals) that has become the focus of public debate in recent years, largely because of the way in which it challenges some of our conceptions of links between maternity and motherhood, reproduction and childrearing. A type of surrogate pregnancy has a long tradition stretching back to ancient times. The Bible gives a very old example.

> Abram's wife Sarai had borne him no child, but she had an Egyptian slave-girl called Hagar. So Sarai said to Abram, 'Listen now! Since Yahweh has kept me from having children, go to my slave girl. Perhaps I shall get children through her.' And Abram took Sarai's advice... And once [Hagar] knew she had conceived, her mistress counted for nothing in her eyes... Sarai accordingly treated her so badly that she ran away from her... Abram was eighty-six years old when Hagar bore him Ishmael. (Jerusalem Bible 1985)

As the Biblical case briefly illustrates, surrogacy arrangements can present a range of different problems. The discussions throughout the text illustrates that advances in reproductive technology have added to the ways in which surrogate pregnancy is achieved. However, while reproductive technologies have given it different expression, the basic idea of having a child for someone else remains the same. The question is: have conditions changed sufficiently to ensure a happier outcome for all concerned?

Activity:

(a) Take a few minutes to jot down your present views on surrogacy. Do you think that it is an acceptable solution to the problem of infertility that some couples have to deal with? Do you think that surrogacy arrangements should be undertaken solely for altruistic reasons: should commercial surrogacy be prohibited?

(b) Would you agree with critics of surrogacy that the process ultimately exploits the women who are contracted into pregnancy and childbearing?

(c) What is your view of the impact on children who are the outcome of this mode of assisted conception?

These activities show that we can formulate very distinct questions about surrogacy. This suggests that it is not one single moral issue but a concept that describes a cluster of issues whose salience depends on the kind of surrogate arrangements that are involved in any given case. In order to get more familiar with some of the key ethical, social and legal issues that emerge in relation to surrogacy read the following paper by Deirdre Madden and Fiona MacCallum.

Contract pregnancy
Fiona MacCallum and Deirdre Madden

Surrogacy, also defined as a 'contract pregnancy', can be described as

> the practice whereby one woman carries a pregnancy for another person(s) as the result of an agreement prior to conception that the child should be handed over to that person after birth. (Brazier Committee Review 1998)

In the traditional method, also known as 'straight' or 'partial' surrogacy, the surrogate mother is inseminated, usually artificially, with the sperm of the commissioning man, making her the genetic as well as gestational mother. Today, however, surrogacy is also described as 'IVF', 'host' or 'full' surrogacy because *in vitro* fertilisation techniques make it possible to implant an embryo in the surrogate mother that is created by the gametes of both the commissioning couple. The role of the surrogate mother in this case then is purely gestational and the child is genetically related to both of the intended parents. On occasions where a donated egg is used, it is possible for the genetic, the gestational and the nurturing mother to be three separate people.

Why do women undergo contract pregnancies?

Contract pregnancies can be private arrangements, perhaps involving a sister or a friend, but more commonly they are mediated either by a fertility clinic (essential for IVF surrogacy) or by agencies specifically established for this purpose. Surrogacy agencies bring together potential mothers with couples who wish to commission a pregnancy. One issue that clinicians or agencies and contracted mothers should discuss is the motivation of these women for carrying a contract pregnancy. There is often concern, whether commercial surrogacy is permitted by law or not, that these women should not be driven solely by financial benefits since this raises the possibility of the exploitation of women who are disadvantaged because of their socio-economic or ethnic status. Additionally, it is feared that a woman who is persuaded to enter a surrogacy arrangement by the prospect of payment may not have considered the extent of the emotional and physical risks entailed and thus not made a fully informed decision. Alternatively, it has been argued that surrogacy is no more exploitative of women than poorly-paid employment in other areas and that the financial rewards are not so great as to lure women into doing something they really do not want to do. In interviews with nineteen women who had carried a contract pregnancy, Blyth (1994) found that only five of them claimed that their main or prime motivating factor was financial. Generally, the concept of remuneration is de-emphasised by both the gestational mothers and the intended parents which allows the idea of the pregnancy being a 'gift' from the mother to the couple to remain, and also fits in with the culturally held belief that children are priceless and not commodities that can be exchanged. However, virtually all of the contracted mothers agreed that it was unreasonable to expect them to give birth and hand over the child with no compensation, even if this only covered reimbursement of expenses. In a study of women taking part in contract pregnancies in the United States, Ragoné (1994) noted that they rarely spent the money they were paid on themselves, but used it to buy things for their families, perhaps as a reward for the disruption to their home and family life during the pregnancy.

Common reasons for becoming a surrogate given by the mothers in Blyth's sample were an awareness of the problems and distress felt by childless couples and the altruistic element of giving the couple what they had always desired. Nine of the women stated that they enjoyed and found easy the experience of pregnancy and childbirth. The women in Ragoné's study used the same reasoning, but she found a conflict between their declared motivations and actual instances. For example, there seemed to be no correlation between the ease of previous pregnancies and the desire to become a surrogate. Women who had experienced difficult conceptions, ectopic pregnancies and even miscarriages were not dissuaded from attempting surrogacy again, even though it put their own health at risk. Since the majority of these women were mothers working in the home, or were in occupations with limited prospects, Ragoné hypothesised that a highly attractive aspect of surrogacy was the opportunity it gave them to transcend their everyday roles, and to participate in something where everybody treated them as special. This was touched on by four of Blyth's subjects who perceived surrogacy as doing something valuable and unusual. One woman said 'I wanted to do something that was out of the ordinary and that made me a little bit special'. There is not necessarily any intrinsic harm in this motivation but problems may arise for the mother when the pregnancy is over, the child is relinquished and she is no longer receiving this special attention.

Activity:

So far, Madden and MacCallum highlight the importance of ascertaining the motivation of women who intend to carry a contract pregnancy. One reason they offer for this, is the concern that if women are motivated solely for financial benefits then these women may become vulnerable to exploitation and the children that they bear would be viewed as commodities rather than persons. Can you think of any other reasons why it is important to ascertain the motivation of possible surrogate mothers? If you were a clinician with responsibility for evaluating the motivation of a potential surrogate mother what criteria would you use to determine whether or not a woman had 'good motivation'?

Discussion

It is useful to pay attention to the way in which surrogacy arrangements are described in different countries and by different authors. In English, for example, a 'surrogate' mother is a 'substitute' mother who acts in the role of mother. Alternatively, the German word for 'motherhood' is 'mutterschaft' and for 'surrogate motherhood' it is 'leihmutterschaft' where the prefix 'leih' means 'rent' or 'loan'. However, in truth, the surrogate is a surrogate for maternity, not motherhood (Lee and Morgan 1989, 2001). It follows that describing the surrogacy as 'full' or 'partial' is incorrect because it is the contribution of the commissioning couple that is partial – they contribute none, one or both sets of gametes – the surrogacy is always 'full' and not divisible according to the genesis of the sperm or egg. Some critics of the term 'surrogacy' have suggested that a more accurate way to recognise the force of the distinctions that the deployment of the terms 'full' and 'partial' are trying to make would be to talk in terms of 'genetic-gestational' and 'gestational' pregnancies (Lee and Morgan 2001). Some describe 'gestational' pregnancies as 'womb leasing' while others argue that the term 'mother' should be reserved solely for the social act of mothering (Mason and McCall Smith 1999: 78, Purdy 2000: 104). Granted the inadequacy of the term, 'surrogacy' is often used synonymously with 'contract pregnancy' and a surrogate mother is increasingly described as a contracting mother or carrying mother. The title of MacCallum and Madden's paper 'Contract Pregnancy', also implies that they are unhappy with the term 'surrogate mother'.

Activity:

Take a few moments to reflect on the terms 'surrogate mother', 'contract pregnancy' and 'carrying mother'. What meaning do the words for surrogate motherhood convey in the language of your country? Can you think of any other ways of describing the surrogacy arrangement? Do various terms connote different value assumptions about this method of 'maternity'? Decide on a term that you think best describes the arrangement.

Returning to the paper, we will see that MacCallum and Madden view the relationship between surrogate mothers and commissioning couples as a very important one.

The relationship between surrogate mothers and commissioning couples

Once a woman has agreed to carry a pregnancy for a couple, it tends to be left to the three of them together to establish a relationship. In situations where the surrogate is previously unknown to the couple, this can be difficult since all those involved are depending on trust between strangers. The founder of one United Kingdom surrogacy agency described it as a 'forced friendship' (Brazier Committee Review 1998). Issues such as the amount of involvement of the couple in the pregnancy and birth, how soon after birth the child will be relinquished and whether contact will continue after the birth, should ideally be discussed at the beginning of the arrangement, preferably before conception, in order to avoid later conflicts.

Surrogacy has been compared to gamete donation in that it introduces a third party into the creation of the family. The intended father must acknowledge the fact that a woman other than his partner will be the mother, at least in the gestational sense, of his child. This goes against the traditional understanding of 'family' in Western society which is seen as established through intercourse and procreation. The fact that the child is a point of connection between the surrogate mother and the father is often underplayed by both mothers and commissioning couples. Ragoné (1994) found that while the role of the father in relation to the surrogate was de-emphasised, the relationship between the commissioning mother and the surrogate was portrayed as a close and special bond and the commissioning mother was very much involved in the pregnancy.

A central concept used by both commissioning and surrogate mothers is the idea of intentionality, i.e., it is the desire of the couple for a child that has brought about the pregnancy even if the intended mother is not physically involved. One surrogate mother stated 'She (the commissioning mother) was emotionally pregnant and I was just physically pregnant'. Similarly, interviews with commissioning parents in the United Kingdom showed that it was expected that commissioning mothers would be present at the birth of the child whereas fathers were present in only two out of twenty cases (Blyth 1995). This extensive involvement of the commissioning mother and the feeling of a shared pregnancy can help the commissioning mother in a traditional or straight surrogacy situation come to terms with the fact that she will not be the genetic mother of the child.

Activity:

Reflect on the quote above from the surrogate mother where she says 'I was just physically pregnant' – the commissioning mother was 'emotionally pregnant'. Does this comment influence your perception of surrogate pregnancy? Do you think the distinction between an emotional and physical pregnancy is accurate to describe the relationship between a commissioning woman and the surrogate? Does the remark tend to minimise the importance and value of the physical labour that goes into pregnancy?

The most widely publicised surrogacy cases have been those where, after birth, the relationship has broken down and the surrogate has decided not to relinquish the child. Although in practice this is very uncommon (it is believed to be around four to five percent), it is a source of concern for commissioning parents (Blyth 1995; Van der Akker 2000). In the United Kingdom, surrogacy contracts are unenforceable although the leading surrogacy agency, Childlessness Overcome Through Surrogacy (COTS) outlines its position in a guide for surrogates: '...if you kept the child. You would have robbed them of all hope they have placed in you' (COTS 1997). The situation is not always as clear as this, since it may be that the couple's behaviour during the pregnancy has made the surrogate mother doubt their suitability to be parents. For example, if a woman agrees to carry a child for a couple to complete their family, and the couple then separate during the pregnancy, she may be concerned about handing the child over to separating or divorcing parents.

Activity:

One case study, which underlines the emphasis that MacCallum and Madden place on the relationship between the carrying woman and the commissioning couple, concerns two couples, whom we'll call the Quinlans and the Simpsons, and baby Emily. This case is a variation of an actual case which was decided upon by an Australian court asked to adjudicate between the carrying mother and the commissioning couple. Take a few moments to read the case and consider Deirdre Madden's commentary on it below.

The Case of Baby Emily

Two couples were friends for a number of years. The Quinlans were unable to conceive, Ms Quinlan had a total hysterectomy arising from ovarian cancer. The couple had an adopted son who was aged three at the time of the trial. The Simpsons had three children of their own aged between three and seven years. Ms Simpson offered to be inseminated with Mr Quinlan's sperm and then to carry the child with a view to handing it over to the Quinlan couple after birth. The arrangement was entirely altruistically motivated.

Baby Emily was born in December 1996 and was taken by the Quinlans to their home a week later. It had been intended that the couples would remain in close contact, particularly between Ms Simpson and the baby. Ms Simpson became frustrated at the level of communication which she perceived as inadequate and she also began to attend grief counselling and a support group for 'relinquishing mothers' . She decided that she could no longer abide by her decision to relinquish the baby and in July 1997 she travelled to the Quinlans home and took the baby from them.

An initial hearing returned the baby to the Quinlans with a contact order in favour of the Simpsons. By the time the trial came to full hearing the baby was one year old and had been living with the Quinlans for most of her young life. The court took the view that Emily should reside with the Simpsons although contact was arranged with the Quinlans and they were to have shared long-term responsibility for her care, welfare and development.

The reasoning behind the judge's decision centred on the possible identity problems which would be faced by Emily during her adolescence and his conclusion that her biological mother, Ms Simpson was better equipped to deal with those problems. Also the judge found that the loss to the child of not growing up with her half-siblings outweighed the loss to her of her relationship with her adopted brother. An appeal from the case by the Quinlans was unsuccessful and Emily was ultimately handed over to the Simpsons in September 1998, when she was almost two years old. (Madden 2000a)

Activity:

This case study highlights what Madden and MacCallum also emphasise: that the relationship between the commissioning couple, specifically, the commissioning mother and the surrogate mother, is key to a successful outcome for all concerned. Take a few moments to reread the case study and consider if anything might have been done at an earlier stage, at the legal level, at the level of reflection (should the couples have sought counselling?) or at the medical level, that might have prevented the breakdown between the Simpsons and the Quinlans from happening.

This case clearly demonstrates the difficulties inherent in surrogate motherhood where the surrogate mother is unable to relinquish custody of the child she has agreed to carry. The problem is further exacerbated due to the close friendship between the surrogate mother's family and that of the commissioning couple. Madden (2000a) focuses on two key issues that arise from this case: motivation and biological ties which we discuss in turn.

Commentary on the Case of baby Emily
Deirdre Madden

Motivation

In this example the question of payment does not arise. Here the surrogate mother was the one who suggested the plan, she already had three children of her own and therefore, arguably, knew her likely response to the hormonal changes brought about by pregnancy and the likely emotional attachment she might feel during the pregnancy and at childbirth. This was an arrangement completely different from what is thought of as commercial surrogacy. It was entered into by all parties with the best of motives and with all the right support structures in place.

The judge did not criticise the couples for having become involved in this arrangement and concentrated instead on the paramount importance of the child's welfare. The court took the view that the arrangement had been entered into with the noblest of motives and that all the adults involved were genuine and well intentioned.

Discussion

While the altruistic motivation of all concerned is underlined in the Emily case, some proponents of contract pregnancy such as COTS in the United Kingdom argue that surrogate mothers should be financially rewarded for their labour and deny that such reward means that these women are selling their babies. Ethical theorist, Laura Purdy, makes the following claim:

If 'selling babies' is not the right description of what is occurring, then how are we to explain what happens when the birth mother hands the child over to others? One plausible suggestion is that she is giving up her parental right to have a relationship with the child. That it is wrong to do this for pay remains to be shown. (Purdy 2000: 108)

Basically, Purdy argues that rather than selling her baby, the contracting woman, is making a biological service available. This biological undertaking earns her a parental right (as opposed to a property right) over the child which Purdy argues the woman ought to be free to exchange for financial gain. She concludes by rejecting the idea that contract pregnancy is only morally acceptable if it is undertaken for altruistic reasons:

People seem to feel much less strongly about the wrongness of such acts [as contract pregnancy] when motivated by altruism; refusing compensation is the only acceptable proof of such altruism. The act is, in any case, socially valuable. Why then must it be motivated by altruistic considerations? We do not frown upon those who provide other socially valuable services even when they do not have the 'right' motive. Nor do we require them to be unpaid. For instance, no one expects physicians, no matter what their motivation, to work for beans. They provide an important service; their motivation is important only to the extent that it affects quality. In general, workers are required to have appropriate skills, not particular motivations. Once again, it seems that there is a different standard for women and for men. (Purdy 2000: 108)

Alternatively, the following extract from the Review of the United Kingdom government established Brazier Committee on Surrogacy argues that any form of payment beyond expenses has to be regarded as a form of 'child purchase':

It was argued by a number of respondents to our questionnaire that surrogacy need not be equated with 'baby-selling', because any fee paid to the surrogate can be regarded as payment for the pregnancy, i.e., payment for her services, not the baby... It is unimaginable that a commissioning couple should enter into a contract that required simply that the surrogate become pregnant and give birth. The contract would have to contain a requirement that in return for the fee the child was handed over to those contracting the pregnancy, with penalties for failure to fulfil this aspect of the agreement. (Brazier Committee Review 1998: 34)

The Committee ultimately recommends that payments to surrogate mothers should cover only 'reasonable' expenses and that agencies involved in establishing surrogate pregnancies should be required to register with the Department of Health and conform to a Code of Practice it would draw up.

Activity:

(a) The case of Emily illustrates that, even when surrogacy arrangements are undertaken with the best of motivation, problems can arise. Stop here for a moment and decide whether or not you agree with the judge that all parties in the case were sincere and well-motivated. Take this opportunity to reflect on what motivates you in your work as a health care professional. Would you describe your motivation as altruistic?

(b) Take some time to consider both Purdy's argument for, and the Brazier Committee's argument against, the payment of surrogate mothers. Decide which of these positions is closest to your own and jot down the kind of surrogacy arrangements (if any) that you think should be permitted.

Now read about the second issue that Madden discusses in relation to the Emily case.

Biological ties

In this case the judge's decision was based almost entirely on the 'welfare of the child' a principle central to many codes of practice and legal systems such as the United Kingdom Human Fertilisation and Embryology Authority (HFEA), the European Group on Ethics and Reproductive Technologies or the United Kingdom and European Union law. The judge focused a lot of attention on the importance of the role of the biological mother as it was felt that she would be in the best position to deal with any identity problems faced by the child in adolescence. In addition, there was very little discussion of the role of Mr Quinlan, the child's biological father. This leads to the assumption that the court felt that the biological mother's role in helping the child to deal with any potential identity problems was necessarily more important than the father's role. I wonder if this assumption is justifiable in all cases and whether or not the judge's decision would have been different if the child were a boy. Or again, if the decision would have been different if the surrogate had been only the gestator of the child and not also the genetic mother.

Sibling ties were also considered to have more importance than the actual emotional bond already formed with an adopted brother. This would seem to ignore the psychological impact on the child (and her brother) of their separation and the difficulty in explaining this loss.

Activity:

(a) Do you think that blood ties are so strong that the emotional attachment of a child to the woman she has known as a mother, and to a brother she has lived with, can be ignored?

(b) The judge placed considerable emphasis on the role of the biological mother in relation to long-term considerations of the child's adjustment. Do you agree with this emphasis? It would mean that in all surrogacy cases the biological mother would be awarded custody. Would this always be in the child's best interests?

(c) What does the idea of 'biological' mother include? Should it be restricted to the woman who gives the egg for the pregnancy? Could the birth mother also be seen as 'biological'? Pregnancy and birth are both biological processes as much as ovulation.

(d) Do you think that this is a good precedent to follow? Consider the implications it would have for adoption situations where the adopted child may also be deemed to have similar identity problems during adolescence.

So far, you have critically reflected on two ethically salient issues in relation to the Emily case and parenthood: motivation and biological ties. Now we will return to MacCallum and Madden's paper which throws light on both the legal and the social aspects of parenthood.

Legal issues in the determination of parenthood

In relation to reproductive technology generally, legislation has been slow to arrive in most jurisdictions. Many countries have chosen to deal with the problems thrown up by these techniques in a piecemeal fashion by adaptation of existing private law doctrines or through professional guidelines for medical practitioners (Shultz 1994). However when it comes to surrogacy, public fears and the 'moral panic' engendered by this practice has ensured, for the most part, a speedy legislative reaction (Morgan 1985).

In any jurisdiction, a decision must be made whether to prohibit surrogacy arrangements or to regulate them. Although ethical and legal difficulties may make them difficult to regulate, experience in many jurisdictions makes it clear that prohibition is not effective. [The case of baby Emily, for example, clearly demonstrates the futility of trying to prohibit surrogacy by legislation. At the time this arrangement was entered into in Australia, legislation was in place which outlawed commercial surrogacy and made all surrogacy agreements unenforceable.] In the United States at least eighteen state legislatures regulate surrogacy, some making it a criminal offence (such as Arizona, Michigan, Utah and Washington), others declaring the contract unenforceable (Nebraska and Indiana). Some permit surrogacy subject to certain conditions such as conditions relating to payment of expenses (e.g., Florida). Some states have dealt with it by judicial decision, e.g., banning the practice entirely as in New Jersey; or in connection with adoption legislation, as in Kentucky. One state has upheld gestational surrogacy as legitimate – California.

In the European context, countries which specifically prohibit surrogacy entirely are Germany, Austria, Sweden and Norway. Other countries, including France, Denmark and the Netherlands prohibit any payment in relation to surrogacy while the United Kingdom permits the payment of reasonable expenses only. Surrogacy also takes place in the absence of any legislative provisions or ethical guidelines in Belgium, Finland, Greece and Ireland.

Activity:

Take a moment to reflect on reasons – social, cultural, religious, historical – why surrogacy is not permitted in many jurisdictions. Can you think of any reasons why policy or law might prohibit surrogacy arrangements? Is it just that surrogacy is considered morally unacceptable, or is it that it is legally 'messy'? In other words, irrespective of whether we consider surrogacy acceptable

from a moral point of view, might there be legal reasons for deciding to prohibit it? What is the legal situation in your own country?

Now return to MacCallum and Madden who discuss the way in
which legal parenthood is determined by the courts.

In any situation in which the surrogate mother decides to
keep the child, a court may have to decide who the legal
mother of the child is: the genetic mother (the woman who
has provided the egg) or the gestational mother (the woman
who carried the child to term) or the commissioning mother
(if she is different from the genetic mother). In most jurisdictions in Europe, Australia and the United States it has been
held that the woman who gives birth to the child is the legal
mother, irrespective of the presence, or lack, of a genetic relationship. For example, the United Kingdom law states:

> The woman who is carrying or has carried a child as a result of the
> placing in her of an embryo or of sperm and eggs, and no other
> woman, is to be treated as the mother of the child. (Human Fertili-
> sation and Embryology (HFE) Act 1990, Section 27, 1)

Section 28 of the Human Fertilisation and Embryology Act
(HFE Act) treats the husband/partner of the carrying woman
as the 'father' of the child if he accepts her surrogate role. In
cases where the carrying woman is single or where her partner
does not accept her role, the child's father is deemed 'unknown'.
This is the case even in situations where the genetic father is
known. In this way, the HFE Act places emphasis on a familial rather than a genetic conception of fatherhood.

In addition to these provisions, Section 30 of the Act
provides for the transfer of legal parental responsibility from
the surrogate parent(s) to the commissioning couple. What has
been called a 'parental order' (as opposed to full adoption procedures) can be granted to a commissioning couple who apply
within the first six months of the child's life. It is granted if

- the child is domiciled with the commissioning couple
- at least one of the two is genetically related to the child
- the surrogate parent has consented to the transfer of
  parental responsibility no earlier than six weeks after the
  birth, and
- no money (other than 'expenses reasonably incurred') has
  exchanged hands.

A different solution to this solomonic dilemma has been put forward in some Californian cases where legal parenthood has been decided on the basis of the parties' intention at the outset of the surrogacy arrangement. In one such case, an argument was put forward that the child should be recognised as having two legal mothers – the surrogate by reason of having given birth to the child, and the commissioning mother on the basis of blood tests which showed her to be the genetic mother (*Johnson v Calvert* 1993). However the court in this case, rejected this argument preferring the claim of the commissioning mother, not on the basis of the genetic link, but rather on the basis of the parties' intention when the surrogacy arrangement began. The majority of the court felt that:

> The parties' aim was to bring Mark's and Crispina's (the commissioning couple) child into the world, not for them to donate a zygote to Anna (the surrogate)... Although the gestative function Anna performed was necessary to bring about the child's birth, it is safe to say that Anna would not have been given the opportunity to gestate or deliver the child had she, prior to implantation of the zygote, manifested her own intent to be the child's mother... She who intended to procreate the child – that is, she who intended to bring about the birth of a child that she intended to raise as her own – is the natural mother under Californian law. (*Johnson v Calvert* 1993: 782)

This has been described as the 'intellectual conception' or the 'but for' test in deciding the issue of parentage. In short, the idea is that without the intention of the commissioning couple, the child would never have come into existence (Douglas 1994).

What are the consequences of focusing on intention in determining parenthood?

- This kind of decision-making is a step closer to the recognition of social as opposed to biological parenthood and some might argue that this does not fit in with traditional norms of a two-parent heterosexual family.
- Because this model of decision-making is unpredictable and subjective, it is difficult to accommodate the welfare and interests of the child within it.
- Proof of intention may not always be easy to find unless written evidence were forthcoming and it implies a willingness to consider children as forms of property that are freely alienable (transferred to others).

It may be argued that refusing to give the surrogate mother parental rights over the child she carried, amounts to a denial of the fundamental importance of her role in nurturing the child for nine months in her womb. Essentially, the surrogate has complete control over the health and safety of the foetus, as it will be affected by her lifestyle, habits, diet and arguably her psychological and emotional health. Her choices as to consumption of alcohol, drugs, cigarettes, nutritional food, vitamin supplements and so on may to a larger or lesser degree have permanent effects on the developing child. The argument, therefore, is that if effects on the child are seen as sufficient to recognise the importance of the genetic input of the commissioning parents, then the surrogate should also be regarded as having made a vital contribution to the developing child.

Discussion

If one considers the gestational role of the carrying woman, who nurtures the foetus in her womb for nine months, is fundamentally important one might favour the position expressed in the United Kingdom, HFE Act 1990, which considers the gestational mother as the legal mother. One reason for taking this position is that the gestational mother has responsibility for giving birth! If the child's welfare is considered, the gestational mother also needs to look after the health and safety of the foetus. She has responsibility of knowing that her lifestyle and choices will have permanent effects on the developing child. The argument would be that if the genetic input of the commissioning parents is recognised because of its effects on the child, then, the contribution of the gestational mother must also be recognised. What is at stake with regard to both genetic and gestational input is the degree to which each of these affect the development of the child. It might well be argued that of the two, the gestational input is, by far, the most significant.

Alternatively, one might favour the intent basis for determining maternity rather than a presumption in favour of the gestational mother. Why? because without the intention of the commissioning couple to bring about the birth of the child and raise him or her as their own, the child would never come into existence. However, MacCallum and Madden draw attention to the following objections that might be raised against the Californian position which considers the 'intentional mother' as the legal mother:

1. It draws us closer to recognising social as opposed to biological parenthood and runs counter to traditional ideas of the family as a two-parent heterosexual unit.
2. The 'intention' to parent is unpredictable and subjective and, therefore, difficult to determine.

Activity:

(a) Consider each of the objections to the Californian position in turn. Do you agree with either or both of them? For example, objection 1 might be viewed as an argument in favour of, rather than against, the Californian definition of parenthood because there are people who would argue that the extension of traditional notions of 'family' beyond the heterosexual unit is a positive thing.

(b) Assume that there are five contributers: genetic mother, genetic father, gestational mother, social/intentional mother, social/intentional father. Which do you think is more 'valuable'? Is one more the 'parent' than others? Do deep rooted assumptions about mothering as biological get pronounced here?

If we return to MacCallum and Madden's paper and to their discussion of some of the social aspects of surrogacy, we will see that they are interested in teasing out the possible consequences of reconceiving 'family' life. They begin by using the concept of 'kinship' to indicate that 'family' does not simply mean 'nuclear family'.

Social issues in the determination of parenthood

An interesting concept that surrogacy invokes is the idea of kinship which has been described as a network of people who are related by blood, marriage or solemn social ties and linked through an enduring bundle of rights and responsibilities. This group partakes in the financial and emotional rearing of children based on a relationship of cooperation, trust and sharing (Kandel 1994). This model may be useful in approaching the problem of custody in a gestational surrogacy case which may be viewed as more a problem of family design. Shared mothering is a distinctive style of child-raising demonstrated, for example, by the Israeli *kibbutzim*, where studies show that the children raised in such an environment showed more advanced intellect, higher ego strength, greater maturity and better adjustments to changes in their environment (Caplan 1968).

In gestational surrogacy, technology makes it possible for a child to have more than one mother. This is seen by many as 'unnatural' as it fails to perpetuate the ideals of the nuclear family. However, since both the gestational and genetic mother both play an essential role in the creation of the child, a legal determination of which of the two is more important is arbitrary. There is no psychological evidence to support the proposition that a child is better or worse off being raised by a woman who is its genetic but not gestational mother and vice versa (Schiff 1995). Perhaps therefore, in cases where there may be a genetic, gestational and social mother, the meaning of motherhood should be expanded to also include non-biologically related adults who have cared for a child and established parental relationships with the child (Bartlett 1984). Although disputes may predictably arise regarding responsibilities and visitation, granting rights to all the parties involved in the child's conception may accord with the wishes of many surrogates who, while they do not want to have primary care for the child, would like to remain involved in the child's life to some extent (Andrews 1989).

Some parents and surrogates maintain a close relationship with visits and phone calls, others limit contact to letters and exchanges of photographs while in some case contact ceases entirely. To some extent, the continuation of contact depends on what the parents plan to tell the child about the contract pregnancy. With other forms of assisted reproduction, such as donor insemination discussed in Chapter 2, concern has been raised that keeping the method of conception a secret will damage family relationships with a consequent negative impact on the psychological adjustment of the child. This view is echoed by the British Medical Association (BMA) Report on Surrogacy which concludes that:

> ... in itself, the fact of being born following a surrogacy arrangement is not sufficient disadvantage to the child to justify refusing the request of the intended parents for assistance with conception. Although little evidence is available, the risk of serious psychological harm to the child is considered low if open acknowledgement is made from an early stage in the child's life. (BMA 1995: 23)

In interviews with twenty commissioning couples, Blyth (1995) found that all of them believed the child should be told the full truth about their origins. The surrogate mothers shared this view. However, whether the parents followed through with this intention is not known. Parents are faced with the decision of how to tell the child and at what age, a dilemma made more difficult by the lack of established narratives for this situation (although, as noted in Chapter 2 this situation is slowly changing).

There is also the possibility that a child, once made aware of the surrogacy arrangement, may decide she wants contact with the surrogate mother even though the parents have decided not to maintain the connection. It could be considered unfair to the child to cut off all ties to the surrogate mother, particularly if she is also the genetic parent. However, one might argue that it is not in the child's best interests to maintain these ties as it prolongs any possible enmity between the parties and reinforces the idea that the child is the subject of a tug of love between them.

Alternatively, it could be argued that it is in the child's best interests to grow up knowing and having contact with the other partners in her conception if this can be done without acrimony. In the Emily case, cited above, the trial judge, in his determination of a custody dispute in favour of a surrogate mother as against the couple with whom the child had been residing almost since birth, paid particular attention to the possibilities for continuing contact between the child and the family, the Quinlans, with whom she had lived. The judge ordered that the family have contact with the child at all such reasonable times as could be agreed between the parties. They were also to have input into the long-term decision-making in respect of the child and to share with the surrogate mother responsibility for her long-term care, welfare and development.

From the point of view of the welfare of children, there is no systematic evidence on the long-term effects of contract

pregnancy on the children thus conceived. Research on families with children conceived by assisted reproduction has so
far found no negative effects on the quality of parenting or on
the well-being of the children, whether or not there is a
genetic link between the mother and the child (Golombok et
al. 1999). However, it is not known how the child will feel
about the unique facets of surrogacy such as the fact that they
have been created in order to be relinquished by their gestational mother, sometimes for financial gain.

Discussion

While MacCallum and Madden note that research so far has
found that there are no negative effects on children conceived
by assisted reproduction, they also acknowledge the possibility that children may feel commodified when they learn of the
facts of their conception and birth through contracted pregnancy. Some indeed argue that contract pregnancy, whatever
its motivation, 'substitutes market norms for some of the
norms of parental love' and that, as a result, children suffer
(Anderson 1993: 171).

> In this practice [contract pregnancy] the mother deliberately
> conceives a child with the intention of giving it up for material advan-
> tage. Her renunciation of parental responsibilities is done not for the
> child's sake, but for her own (and if altruism is a motive, for the sake
> of the intended parents)... One can question whether the sale of chil-
> dren is as harmless as proponents contend. Would it be any wonder
> if a child born of a surrogacy arrangement feared resale by parents
> who treated the ties between a mother and her children as properly
> loosened by monetary incentive? (Anderson 1993: 171–2)

Activity:

Recall the discussion in Chapter Two about possible reactions
from children conceived through donor insemination when they
discover their genetic origins. In working that chapter you may
have come to some conclusion about the reaction of children
discovering they were born through DI. As mentioned above,
some research on surrogate pregnancies indicates that children
may feel commodified when they learn of their 'contractual' pregnancy through surrogacy.

(a) Do you think such research indicators constitute an objection to surrogacy? Are the child's best interests threatened in the arrangement?

(b) How important, in legal and ethical terms is the 'reaction' of the child?

(c) Is the language of 'sale of children' true to the situation of surrogate pregnancies which are contracted because of a desire to have a child?

It could be objected that what is unique about surrogacy is not the fact that the surrogate mother relinquishes the child, but rather, the fact that she undertakes pregnancy and labour in order to fulfill the deep desire of another person to have a parental relationship with a child. Purdy compares the surrogacy situation with other situations in which conception is achieved:

> Considering the sorts of reasons why parents have children, it is hard to see why the idea that one was conceived in order to provide a desperately-wanted child to another is thought to be problematic. One might well prefer that to the idea that one was an 'accident', adopted, born because contraception or abortion were not available, conceived to cement a failing marriage, to continue a family line, to qualify for welfare aid, to sex-balance a family, or as an experiment in child-rearing. Surely what matters for a child's well-being in the end is whether it is being raised in a loving, intelligent environment. (Purdy 2000: 108–9)

Activity:

(a) Earlier in this chapter we considered the relative importance of biological ties to family relationships. In relation to baby Emily, the judge decided that biological ties, specifically the genetic and gestational origins of baby Emily, should be privileged in decisions regarding her custody. This might be a further opportunity to revisit the arguments given in chapter 2 on DI for discussion of the relative significance of social father and biological father. Do cases and discussions show partiality to the value of the maternal biological link as in the Emily case and less valuing of the paternal biological link as in the DI discussion?

Jot down two or three reasons why people might argue that social, emotional and psychological ties are also very significant in decisions concerning the welfare and well-being of children. Now that you have reflected on the significance of biological and social ties, revisit the Emily case and re-evaluate the judge's decision.

(b) Do you see any advantages or disadvantages in having more than one mother? Or father? What kind of kinship structures have you yourself experienced, or met with in your professional life?

(c) Would you agree with Purdy that the circumstances of many people's conception and birth are less than ideal and that the well-being of a child is far more dependent on his or her environment than his or her origins?

In what follows we have compiled a table of some of the arguments for and against surrogacy that have been presented in this chapter. Take a few minutes to read the table and see if you can construct more arguments of your own or, in fact, whether you would take issue with some of the arguments offered in the table.

Arguments for and against Surrogacy

For	Against
✔ surrogacy is the only chance for some couples to have children	✗ surrogacy interferes with the natural state of things
✔ it challenges assumptions about biological/social forms of parenting and offers single people and gay couples a chance of happiness or fulfilment	✗ it threatens marital life and traditional family structures
✔ it is no more exploitative of women than poorly-paid employment	✗ it exploits women who are disadvantaged because of their socio-economic or ethnic status

✔ children are not being bought, rather a woman is being paid for her gestational services, as are other professionals involved in the conception and birth of the child

✘ it distorts the relationship between mother and child and it commodifies reproduction and childbirth to the extent that children are being bought and sold as property

✔ there is a very low rate of problem cases: around four to five percent and there is no evidence of significant harm to the children involved

✘ when surrogacy arrangements break down everyone suffers, especially the child involved

✔ ?

✘ ?

Summary

This chapter focused on some of the key ethical, social and legal issues that surrogacy raises:

• we discussed the social and legal significance of the relationship between the commissioning parents and the surrogate mother
• we saw that the insistence on prioritising altruism in motives of a surrogate woman raised questions about why altruism is necessarily better than a remunerative transaction
• we noted that the debate is clearly not resolved about the relative importance of biological, emotional and social ties to a child
• we touched on the argument that the practice of surrogacy is unethical because it commercialises reproduction and childbirth and treats children as a form of property to be bought and sold
• finally, the question of whether surrogacy is exploitative of women was not answered with a resounding yes or no.

In Chapter 5 we will focus in greater detail on the commercialisation of particular reproductive processes and consider the claim that women should benefit from their labour in pregnancy.

– 4 –

Embryo Research and Stem Cell Therapy

Objectives

At the end of this chapter you will have an understanding of:

- different views on the status of the human embryo
- the implications of these views for accepting or rejecting a ban on embryo research
- the reasons offered for and against a moratorium on stem cell research
- the slippery slope argument and how it applies to embryonic research

The next two chapters of this workbook deal with one of the most controversial topics in ethical debate: embryo research. Chapter 4 addresses one of the central ethical issues in this research: the development of what are called 'stem cell lines' from human embryonic tissue. The use of IVF in assisted procreation has made it possible for stem cell lines to be taken from embryos developed in the IVF process. Embryo research and stem cell therapy developed largely from the progress in IVF work and so offers an opportunity to consider the ethics of these possibilities. Chapter 5 deals with a separate but related issue: whether or not pregnant women should be considered to be the owners, in some sense, of the embryonic tissue (and the stem cell lines derived from it) that they have laboured to produce.

Perhaps some of you think that the topic of embryo research is irrelevant to health care practitioners, for example, family doctors or general hospital nurses. If a general practitioner claimed that embryo research is of no 'practical relevance' to family medicine, how would you respond? As you read through the chapter and consider the case study of the Nash couple's decision you may have a clearer basis for answering that question.

Human embryonic 'stem cells' have become a valued commodity for purposes of therapeutic research but to understand how this is so, we need to ask: what are stem cells?

What are Stem Cells?

Stem cells are long-living cells that have varying capacities to differentiate into specialised tissues and parts of the body such as organs, blood, nerves, skin or bone.

A *stem cell line* is a group of stem cells that is capable of sustaining continuous, long-term growth in a laboratory culture.

Stem cells may be

- *totipotent:* able to develop into all the different types of cells needed for a complete, functioning organism
- *pluripotent:* able to develop into many types of tissue but not into a functioning organism
- *multipotent:* able to give rise to a limited number of tissue types

All three kinds of cells are present in early human development, while multipotent stem cells can also be derived from children and adults.

If this seems like asking you to step into deep scientific waters, be assured that, as you work through this chapter, you will read further clarifications on stem cells, where we can procure them, and various proposals for their use in research and treatment of disease. But before we get into the detail of the ethics of stem cell research and therapy, read about a well-known case concerning stem cells that appeared in the media throughout the world in 2000. The case involves the transplantation of multipotent stem cells from the umbilical cord of one child in order to save the life of another. The case has engendered widespread discussion about the ethical issues that arise. It is believed to be the first known instance where preimplantation genetics was used both to screen for a disease and to ensure a tissue donor match in a sibling.

As you reflect on the case, think about the ethical dilemmas it poses and also think about what it tells you about stem cell research in general, its importance and its drawbacks.

The Case of Molly and Adam Nash

Six-year old Molly Nash was born with Fanconi anaemia, a rare genetic defect that prevents her body from generating bone marrow, that produces blood cells. Doctors estimated that her life expectancy was only about seven years and that she had already begun exhibiting symptoms of the onset of leukaemia. Definitive treatment of the disorder relies on reconstituting the patient's bone marrow. Bone marrow reconstitution can be accomplished via bone marrow transplantation or umbilical stem cell transplantation. Transplanted umbilical stem cells have the capacity to migrate to the recipient's bone marrow, take up residence there and differentiate into immune and blood cell prescursors. The procedure is both painless and harmless for the donor since the cells are taken from the umbilical cord which would otherwise be discarded.

Molly's only chance of survival was a bone marrow transplant. It was estimated that a transplant from an unrelated donor would provide a thirty-one percent chance of survival, whereas a transplant from a tissue matched sibling would raise the chance of survival to between eighty and ninety percent. The parents, Lisa and Jack Nash, had planned on having other children but were hesitant as they both carried the gene for Fanconi's anaemia and were advised that they had a twenty-five percent chance of conceiving another affected child if they used conventional means of reproduction. They decided to have IVF treatment and to use preimplantation genetic diagnosis (PGD) to screen the embryos with two aims in mind:

- to eliminate embryos with the genetic defect that causes Fanconi's anaemia.
- to eliminate the embryo(s) that do not have a tissue match for Molly.

Only two of the fifteen embryos derived from the IVF process fulfilled the aims here: that is, were perfect tissue matches *and* free of the disease. However, only one embryo survived the implant procedure. Some of the embryos not implanted were disease free but, since they were not tissue matched for Molly, they were not used in this procedure and were subsequently destroyed.

The single embryo that survived the implanting process produced a little boy, Adam. Blood cells taken from his umbilical cord were transplanted into Molly in order to provide Molly with the necessary stem cells to build new bone marrow. Stem cells in the cord blood migrated to the hollow areas of her bones, where they lodged and began producing healthy blood cells. Molly's blood counts have increased steadily since the transplant, indicating bone marrow recovery. Her body has begun for the first time in years to manufacture platelets and neutrophils, a white blood cell critical to fighting bacteria. There will be a year of observation for rejection and infection but Molly's prognosis is so positive that doctors plan to allow her to return home four months earlier than planned. The transplant will eventually eradicate the blood disorder in Molly, but she may be more prone to certain types of cancers in adulthood. (Madden 2000b)

Discussion

In her commentary on the case, Madden (2000b) notes that this case may be seen as ethically unproblematic if the parents intended to have more children anyway and Adam's birth simply provided, by happy coincidence, the means by which their daughter's life could be saved. However, if Adam was conceived just for this purpose it may be argued that conceiving a child solely in order to benefit another person is unethical in that it treats the child only as a means to another end. Respect for a child's own person would seem, on this view, to go counter to such instrumental use. In brief, if Adam was conceived solely to treat Molly, then Madden argues, the procedure wronged Adam. But we need to take a closer look at what is meant by the claim that an action is unethical if it involves the instrumentalisation of someone or something. Consider the facts of the case, might Adam be considered a means as well as a value in himself?

What kinds of things are good or valuable?

In trying to understand what is meant when someone is concerned that a child like Adam is being 'instrumentalised', we might begin by asking 'what kinds of things are good or valuable?' To respond we can cite two kinds of goods often discussed in ethics:

1) purely intrinsic goods (of which simple joys are an example)
2) purely instrumental goods (of which the practice of making money is an example)

We consider some things good or worthy of desire (desirable) in themselves (intrinsic goods) just for what they are independently of what other goods they might lead to. We can then say of other kinds of things that they are 'good', valuable or desired because they promise to lead to or produce other good things. These goods or values that are seen as 'means to other ends' are called instrumental goods. So, instrumental goods are worthy of desire because they are seen as effective means of attaining other goods that are often intrinsic goods – valued in their own right alone. Socrates mentions two instrumental values: money and medicine. Medicine is an instrumental good in that it is not normally valued for its own sake – as something sought purely for itself. Rather, with the example of medicine, we can ask: What is medicine for? The answer is, 'to promote health'. But notice here, Socrates thinks health is an example of an intrinsic good *and* an instrumental good: health is good in itself but it is also good for other things as well such as creative activity, achievement in sports, etc. Money is perhaps solely an instrumental good or value. Most of us don't value money for its own sake but we tend to value money for what it can buy. Money can buy entertainment, shelter, clothing, food, musical instruments, etc.

Take the example of reading this chapter on embryo research. At the moment you might say that you are using this text instrumentally, as a means of learning about the ethics of NRTs. You might say that this text has 'instrumental value' to you. But reading this text *also* offers intrinsic value since knowledge can be considered a good or value just in itself without reference to any other end. So, we can see a third kind of good where some things have a combined value: both intrinsic and instrumental. As mentioned above, Socrates gives the example of 'health' and considers it a 'combination good'. A combination good then is exemplified in such things as sight, health and knowledge – all both good in themselves (or intrinsically) should one choose to luxuriate in knowledge without further goals or objectives in mind for using knowledge. But knowledge and health are also goods as means to further goods. What is considered of intrinsic or non-instrumental value is often a matter of degree that is determined within a context. Going back to our example again, while the text may be a means to your learning ethics, you may consider that learning ethics is a means to your becoming a competent professional or wise human being. Relative to your personal and professional goals, the activity of learning ethics has instrumental value.

Now let's try and apply this distinction to the Nash case. Madden's position shows ethical unease about the Nash decision because she claims that Adam might be instrumentalised here if his birth is being treated not as a good thing in itself but rather solely as a means to achieve the health of his sister, Molly. On Madden's interpretation, this way of treating Adam and viewing him as instrumental would be ethically dubious.

Activity:

(a) Re-read the Nash case and reflect on the distinctions made above about intrinsic, instrumental and combined values or goods. What do you think about Madden's ethical concerns here?

Is Adam being used or treated solely as a means to a therapeutic end for Molly?

Is Adam a combined good?

(b) Do you think all instrumental use of human persons is ethically suspect? If so, why? Try to think of examples of persons being means to some other good.

When we ask others to donate blood for a depleting blood bank, we are hoping they will serve as a means to help others. We assume they will choose to give blood, whereas with Adam he had no choice in the matter. Does this element of choice make a difference to your view of the ethics of the Nash case?

What did you decide? Immanuel Kant, an eighteenth century moralist would argue that it is wrong ever to treat a human person solely as a means to an end (Kant (1991[1785]). He puts it this way: 'Act in such a way that you alway treat humanity, whether in your own person or in the person of any other, never simply as a means, but always at the same time as an end.' But notice that Kant says 'simply' as a means and always treat persons 'at the same time' as an end. Kant is holding a position here that 'combined goods' are genuinely possible and one and the same good can be 'intrinsic and instrumental'.

If we have children because we hope very much to have companionship or care in later years, is this treating children simply as means to our ends? Is this instrumentalising children in an ethically negative way? It seems evident that women or couples have children for all kinds of reasons – and for no reasons at all. Madden is concerned about treating Adam 'instrumentally'. However, at the same time, it could be argued that Adam's umbilical cord, if not therapeutically put to use, will be incinerated or passed on to a research company. Moreover, Adam, as donor, is not harmed in giving this umbilical cell help to his sister.

Do you think the facts of the case suggest that Adam can accurately be viewed as a combined good – valued for his own sake and, instrumentally, as a gift to Molly?

Activity:

(a) In the course of securing a 'suitable' embryo for implantation, fifteen embryos were grown in vitro, thirteen of these were destroyed and one failed to implant.

Were the destroyed embryos used in a solely instrumental way here as means to achieve the 'right' embryo for Molly? Was value accorded the discarded embryos?

(b) What is your reaction to the destruction of a number of preimplantation embryos, so that a healthy child is born? You might stop for a few moments and think about what an embryo means to you. Would you describe an embryo as a person, a human being, a potential human being, a cluster of cells or tissues, or??

The Nash case is a case study about the prospects for using embryonic stem cells from the IVF procedure. The use in this case is therapeutic and directed toward greatly improved health for Molly. You will need to assess whether you can ethically justify the process and means used to achieve this desirable therapeutic end.

Part of the ongoing ethical debate about stem cell research has to do with the sources of stem cells. Where do they come from? The discussion below explains three sources where stem cells can be obtained. These three sources differ in a number of respects and also offer varying degrees of 'promise' in the therapeutic application of stem cells. You might wish to take notes as you proceed here and jot down some ethical questions that surface for you in each of the three methods of deriving stem cells.

What are the sources of embryonic stem cells?

Totipotent and pluripotent cells which can give rise to a multiplicity of cell types are classified as embryonic stem (ES) cells. Those cells which can give rise to gametes – sperm or eggs – are called embryonic germ (EG) cells. They can be sourced in three different ways, from:

spare IVF embryos,
embryos created especially for research, and
cells from aborted foetuses.

1. ES cells from spare (supernumerary) IVF-embryos created for infertility treatment but no longer needed for this purpose
2. ES cells from embryos created for the sole purpose of research from donated gametes that are fertilised *in vitro*. Research embryos can also be derived from what is known as **somatic cell nuclear transfer** (SCNT). This is the most 'revolutionary' and controversial procedure, which is also defined as **'therapeutic cloning'**. It involves the creation of an embryo by inserting a nucleus derived from the somatic cell of a patient's own body into an enucleated egg-cell
3. EG cells from foetuses aborted 5 to 9 weeks after fertilisation, specifically, from the region that is destined to develop into the sperm or eggs

The use, creation and destruction of embryos in ES stem cell research is justified by its proponents because of the potential of stem cells to offer immense health benefits. Up to now, research that involved destroying embryos, if allowed, was limited to research on reproduction, contraception or congenital diseases. With stem cell research, however, the scope is very much widened. It is argued that cells derived from embryos will eventually provide a permanent resource that can be used to generate replacement cells and tissues to treat any number of diseases such as leukaemia, Alzheimer's and Parkinson's or injuries such as spinal cord injury or burns. In short, it is widely agreed that 'cell-replacement therapy' (the replacement of diseased cells with ES cells) will revolutionise current ways of diagnosing and treating human illness.

While research on stem cells from each of the three sources raises a cluster of serious ethical, social and public policy issues, so far public debate about them has moved far more slowly than the scientists who are engaged in stem cell research. And what debate there has been has not resulted in any kind of consensus on the way forward. While ethicists ponder, corporate funded ES cell projects continue to develop.

Activity:

(a) Note the different sources of embryonic cells:

   (i) ES cells are derived from supernumerary IVF-embryos, approximately five to fourteen days after fertilisation.

   (ii) ES cells are also derived from embryos that are created through IVF for that specific purpose. On both of these methods, (i) and (ii), the embryo is effectively destroyed by the removal of the outer layer of the blastocyst.

   (iii) EG cells are derived from the reproductive cells of aborted foetuses and although the cells are still alive, the foetuses are dead before research starts.

Make a list of the ethical differences (if any) that you can identify between these several procedures.

(b) Is there a debate on stem cell research in your country? Try and give reasons for the interest or lack of interest in this issue.

The status of the embryo

Because there are widespread differences on the issue of stem cell research, we will briefly outline the current state of the debate. Questions about the acceptability of research, in general, using human embryos have prompted a number of governmental ethical committees throughout Europe and elsewhere to examine it. [See for example, (Canada) Royal Commission on New Reproductive Technologies 1993; (France) Comité Consultatif National d'éthique pour les sciences de la vie et de la santé 1993; (United States) National Institutes of Health 1994; (United Kingdom) Nuffield Council on Bioethics 2000.] For purposes of encouraging you to think about how you evaluate the human embryo, we examine an extract from a Report of the Health Council of the Netherlands Committee which offers a summary of three different positions on the status of the embryo *in vitro* and the implications of these positions for the use of such embryos as a source of ES cells.

The moral status of human beings is connected with their personhood. As persons, they deserve respect and, where necessary, protection. The question is whether and to what degree human embryos *in vitro* are persons and worthy of protection. The brief responses provided to this question can be divided into three groups, according to the underlying view of the relationship between the moral status of post-natal human beings and the moral status of human embryos.

1. Equally worthy of protection

According to some, the moral status humans have as humans (persons) can also be attributed to the same degree to human embryos *in vitro*. On this view, the embryo has intrinsic value, is no less worthy of protection than a child or an adult human being, and should be treated accordingly.

A variant of this view is based on the concept of personhood as used in Roman Catholic teachings. According to that concept, personhood is seen as given, not with any human faculty, but with the essence of humanity as it appears in the unity of body and soul. Although this view leaves space for uncertainty about whether or not it is a person at any time during its development, this does not constitute moral leeway: we cannot risk killing a person.

Another variant of the 'equal worth' view assumes the 'modern' concept of personhood, which is defined in terms of certain human faculties (which in any case include a form of self-awareness). According to this view, the fact that human embryos do not fulfil criteria for personhood given with that definition is not conclusive. One conclusion is that they can be considered as 'potential persons'. Potential persons would have the same moral status as persons and would therefore be worthy of the same degree of protection.

2. Not worthy of protection (but with a symbolic value)

Others believe that human embryos *in vitro* have no moral status that makes them worthy of protection. Proponents of this view do not regard the capability of developing into a human being as the actual purpose of an embryo but only as a possibility, which moreover depends on whether or not it will be implanted and allowed to develop. According to this view, there is no reason to treat the embryo *in vitro* differently from the gametes from which it emerges.

Having 'no moral status' need not mean that human embryos *in vitro* may be subjected to every form of treatment. Some ways of working with human embryos specifically in research may be opposed for other reasons, including the 'symbolic value' that human embryos (a beginning form of human life) actually have in our society. So conceived, it is not for the sake of the embryo itself that it ought to be spared but for the sake of the community in which it has acquired social significance.

3. Relative worthiness of protection

A characteristic of a third group of views is that moral relevance is attached to both the difference (the embryo is not a human being) and the continuity (a human being could develop from the embryo). This represents a position between the two views discussed above. On the grounds of its human origin and its potential to develop into a human being, the embryo *in vitro* has an intrinsic value according to this understanding and therefore is worthy of protection. On the other hand, it is stressed that the embryo's moral status is not equal to that of human beings. The embryo's worthiness of protection is therefore regarded as relatively limited.

Often this is augmented with the notion that the moral status of human embryos and foetuses increases with their development. In different variants of this gradualist view, various 'moments' in embryonic (or foetal) development are referred to as a transition to a stage in which the embryo would be more worthy of protection, e.g., the appearance of the primitive streak or the commencement of brain activity. (Adapted from the Health Council of the Netherlands Committee on In vitro fertilisation 1998)

Discussion

The Netherlands' Health Council concludes that only those adopting the first position described above would reject embryo research absolutely and that the other two positions would not deem embryo research morally unacceptable in advance. The Council argues, however, that insofar as the embryo is given some kind of value, any form of research using human embryos requires further justification:

> The fact that it might lead to greater knowledge or that useful applications are conceivable is not sufficient; it has to be clear that the interests served by the research are also morally more important than the value attributed to the embryo *in vitro*. (Health Council of the Netherlands 1998/08E: 52)

The Dutch demand for justification of embryo research finds echoes in other jurisdictions which permit it or are considering doing so. In fact, in any of these jurisdictions, it is stipulated that research be carried out only on *in vitro* embryos up to fourteen days after fertilisation (prior to the appearance of the primitive streak) and within the terms of strict codes of practice. The fourteen-day limit would seem to reflect a gradualist view of the embryo as spelled out in 'relative worthiness' position three given above.

Activity:

(a) Take a few moments to consider each of the three positions outlined by the Dutch Health Council that the embryo is (1) equally worthy, (2) of symbolic worth, (3) relatively worthy of protection. Decide which of these three comes closest to your own position on the moral status of the human embryo. Jot down two or three reasons why you choose this particular position.

(b) Now think about the implications of your position on embryonic research.

● If you consider that a human embryo has an intrinsic value or is equally worthy of protection to that of a child or adult, then, to be consistent, you must agree to an outright moratorium or ban on embryo research.

If you (or your own country) hold this 'intrinsic value' position, must you also refrain from using the benefits of stem cell research altogether?

● If you consider that the embryo has no intrinsic value but only symbolic value to a community, then you may allow that using it in a solely instrumental way as a means to other goods is ethically permitted. (You may still disagree with embryo research, but you must base your disagreement on reasons other than the status of the embryo).

● Finally, if you consider that the embryo has relative value, you will have to weigh the value of the embryo against the potential therapeutic benefit of carrying out research out on it. The therapeutic benefits are for living human beings but also future human beings. How would you attempt this weighing up of the value of the embryo against therapeutic benefits?

When reflecting on the three different positions on the status of the human embryo, a sense of frustration is voiced by some students of bioethics. When invited to clarify the source of their frustration, students say that the frustration arises when one can't seem to get any definitive answer to a question they consider important and that they think (rightly or wrongly) should have a clear, distinct and unambiguous answer. Such an important question concerns the status of the human embyo. However, why do you suppose that, in all the many years of considering that question, scholars, laypersons and clergy have not been able to resolve differences of viewpoint about it? Is it perhaps that this question cannot be resolved in a determinate way to the intellectual or ethical satisfaction of everyone because the answer is not solely dependent on factual information about the physiology of the embryo? A room of embryologists might all agree on the physiological facts about the developing embryo and yet disagree about the normative aspects, that is, how the unborn embryo should be treated, whether it is a person, and whether it should receive unrestricted protection from a particular time in gestation.

Why do you think this disagreement persists in spite of agreement on all the 'facts of the matter'? What do you think accounts for your 'preferred position' on the status of the embryo? In considering the question: what is the status of the human embryo, we are faced with a question that is made 'determinate' or given some precise answer only with reference to differences about larger values about the 'good' in life, assumptions about the world and whether a deity is part of that world or not. Perhaps the lack of an answer to this question leaves some dissatisfied because they think it's important enough to be able to get agreement among everyone. But, ironically, many of the diverse 'answers' offered to the most important questions in existence reflect different world-views that are often of a religious nature and involve deeply held commitments with moral consequences.

Activity:

Can you give examples of some other questions where disagreements about the answer reflect much larger differences in world-views and beliefs about human nature? What about capital punishment? Obligations to feed the starving people in the world? The abolition of inheritance as a requirement of justice?

The paper that follows pursues the implications of different assessments of the human embryo. These ideas come from Guido de Wert and Ron Berghmans (2000a). Their paper covers questions on ethics and policy concerning human ·embryonic stem cells and their use in research and therapy. The central question they address is: Is it morally allowed to use and/or produce living human embryos for the preparation of embryonic stem cells? In brief, the authors take it that the embryo has a relative moral value (the third approach to the status of the embryo outlined by the Dutch Council) and that, as such, its instrumental use is ethically acceptable at least in principle. What is at stake for the authors is the extent of this use: the limits and conditions that must be drawn in the light of the relative value of the embryo coupled with the importance of the goals of embryo research.

Human embryonic stem cells: ethics and policy

Guido de Wert and Ron Berghmans

The dominant view is that the instrumental use of preimplantation embryos (in view of their relative moral value) can be justified if specific conditions are met. The international debate focuses on the specific conditions for embryo-research, in particular with regard to the goals of the research and the question whether or not only spare IVF-embryos may be used, or embryos exclusively created for research purposes.

It is generally agreed that the research must serve an important goal; sometimes this is formulated in terms of 'important health interest'. Difference of opinion concerns the way in which this condition should be made operational. In a number of countries embryo-research is restricted to research that is related to human reproduction. Internationally, such a limitation is growingly disputed. The use of ES cells for research into the possibility of cell-replacement therapy [therapy using stem cells to replace existing diseased cells] in this context operates as a catalyst.

Is such research into ES cells morally acceptable? What are the ethical objections, and are these convincing? To start answering these questions, we first restrict ourselves to spare IVF embryos as a source of ES cells. Possible objections are connected to the slippery slope argument and the possibility of other alternatives to spare and specially created embryos as sources of stem cells. We'll call the two objections the 'slippery slope objection' and the 'other options' challenge. We'll begin by explaining the slippery slope argument.

Slippery slope argument

Should research into ES cells in the context of cell replacement therapy be considered unacceptable because this may ultimately lead to the use of ES cells in the context of germ line gene therapy and/or therapeutic cloning?

Discussion

Fears of a slippery slope usually translate into a caution to avoid taking any first step that might put us in a position of a 'slide' towards a position that would be ethically and socially intolerable.

The logic of the slippery slope argument claims that we ought to be extremely cautious about doing x (e.g., cell replacement therapy) which one admits is a good and therapeutically very beneficial human treatment. But the extreme caution which sometimes leads to calls for moratoriums on certain procedures is based on our belief that doing x will, in all probability, lead to y or z. And to understand the slippery slope here, y or z are clearly not places we wish to go! So, we judge on the basis of our moral beliefs that y or z are procedures or types of research that are repugnant or fraught with problems, for example, concerning respect for human beings, rights of individuals, etc. We think there are strong empirical grounds or reasons to believe y and z (y or z) will happen if we do x. So, we conclude that y and z would be the bottom of a dangerous ethical slippery slope and, had we not stepped on top of the slope by choosing x, we could avoid y and z. So notice the slippery slope worry: by doing some good things in research (x) coupled with knowing how research activities have a certain 'inevitability of development' about them, one judges that unacceptable procedures (y or z) are very likely to be conducted and even accepted. Reject a relative good x, and we can be secure in believing that y and z, undesirable occurrences, will not happen. Notice that people might disagree about the dangers of a slippery slope in any particular situation for one of two reasons at least: a) because they disagree about whether or not y or z are abhorrent or ethically unacceptable. That is, often times, in discussions the wrongness of a likely unwanted outcome is taken for granted rather than argued for or, b) whether the anticipated occurrence of y or z is not at all necessary or inevitable.

Now, de Wert and Berghmans do not think that a slippery-slope objection is convincing. Their reason is that, even if germ line gene therapy and therapeutic cloning would be unacceptable (which is not self-evident) it does not necessarily follow that the use of ES cells for cell-replacement therapy is unacceptable. However, the philosopher, David Lamb in *Down the Slippery Slope* (1988) does not as readily discount the significance of 'slippery slope' discourse. He might agree that there is no necessity in y or z happening if x occurs. But, Lamb argues that the slippery slope metaphor is not simply an emotive confusion about progress in a variety of medico-ethical areas. Instead Lamb thinks the concept of a 'slippery slope' might function very constructively as a cautionary note to take vigilance, to tread slowly, thoughtfully and responsibly in the technologically powered transitions from chance to choice which we discussed in Chapter 1.

Activity:
Germ line gene therapy and therapeutic cloning are the two procedures introduced by de Wert and Berghmans to illustrate the slippery slope argument against ES cells research. We will take some time to reflect on the concerns that many people have with each of them.

1. Germ line therapy

First of all, what is germ line therapy? Let's take it one definition at a time:

Germ and somatic cells
There are two classes of cells found in the human body: **somatic cells** and **germ cells**. Germ cells are found in the ovaries of a female and the testes of a male. They give rise to ova and sperm respectively. Somatic cells are all the other cells in the body.

Somatic cell therapy
- typically involves giving a person healthy DNA to override the effects of their own malfunctioning DNA
- is not considered to be very different (ethically) from giving a person a blood transfusion or organ transplant

> *Germ line therapy*
> - typically involves inserting a healthy gene into diseased germ cells
> - is considered too risky at present because of the fear of damage to existing genes
> - is unlike somatic therapy because changes to somatic cells cannot be passed on to children and to future generations, while changes to germ cells can (adapted from Reiss 2001: 24–26)

While somatic gene therapy is viewed as a great benefit to human health, germ line gene therapy is considered unacceptable because of the influence it will have on future generations. The European Convention on Human Rights and Biomedicine, for example, takes the following position:

> An intervention seeking to modify the human genome may only be undertaken for preventive, diagnostic or therapeutic purposes and only if its aim is not to introduce any modification in the genome of any descendants. (Article 12, Council of Europe 1996)

Activity:

Do you agree with the position on germ line gene therapy expressed in Article 12 of the Convention? Compare it with the following position adopted by Reiss:

> Despite the difficulties of distinguishing in all cases, genetic engineering to correct faults (such as cystic fibrosis, hemophilia, or cancers) from genetic engineering to enhance traits (such as intelligence, creativity, athletic prowess, or musical ability), the best way forward may be to ban germ line therapy intended only to enhance traits, at least until many years of informed debate have taken place. (Reiss 2001: 26)

What is your opinion of Reiss' moratorium? Do you think that the benefits of germ line therapy outweigh the risks?

2. Therapeutic cloning

Concerns that ES cell research will lead to therapeutic cloning, is one of the slippery slope arguments that de Wert and Berghmans raise. We have already mentioned somatic cell nuclear transfer (SCNT), otherwise called therapeutic cloning, in passing. If you remember, at the start of this chapter we indicated that there are three sources of ES cells: (1) spare IVF-embryos; (2) created embryos; and, (3) aborted foetuses. With regard to source (2), embryos can be created in two ways, from donated gametes that are fertilised *in vitro* and from SCNT. We will now look at this issue in more detail.

First of all we will distinguish between 'therapeutic cloning' and 'reproductive cloning'.

SCNT or *therapeutic cloning* involves the creation of an embryo by transferring a somatic cell nucleus of a patient to an enucleated egg in order to develop an embryo from which to derive identical stem cells for therapeutic purposes. The embryo thus created does not end up as a living baby because the stem cells are removed at the blastocyst stage, five to six days after the transference.

Reproductive cloning would occur if the embryo created through SCNT were successfully implanted in a womb and developed to full term. The baby, subsequently born, would be genetically identical with the individual from whose somatic cell she originated.

We will now examine some of the ethical arguments for and against therapeutic cloning. Let's first look at de Wert's and Berghmans' views on this issue.

A **consequentialist** objection (fashioned as a slippery-slope argument) is that therapeutic cloning will inevitably lead to reproductive cloning. This position is not convincing; if reproductive cloning is unacceptable (the debate about this issue is still going on), then it is reasonable to prohibit this specific technology but not to ban other applications of cloning.

If, as some argue, it is acceptable to create embryos for research aiming at the improvement of NRT (freezing of egg cells; *in vitro* maturation of egg cells, etc.), then they would argue that it is inconsistent to reject therapeutic cloning in advance as being very likely to lead to reproductive cloning. Maybe even some opponents of creating embryos for the improvement of NRT can conditionally accept therapeutic cloning because of the important health interests of patients.

Discussion

For de Wert and Berghmans, therapeutic cloning cannot be ruled out on the basis that it leads to reproductive cloning. Their main reasons are that:

- there is no inevitability that therapeutic cloning will lead to reproductive cloning
- the ethical debate on reproductive cloning is not closed and settled in the negative so it would be premature to rule it out in advance
- anyone who considers that the creation of embryos (presumably from gametes fertilised *in vitro*) is justified cannot, without being inconsistent, object to therapeutic cloning

A similar argument defending therapeutic cloning comes from the UK Nuffield Council on Bioethics (2000). According to Dickenson (2000) it offers a **utilitarian** defence of stem cell research generally and comes to the conclusion that 'the removal and cultivation of cells from a donated embryo does not indicate lack of respect for the embryo' (p.1). The Council looks at a number of different methods of obtaining stem cells and claims that it is ethically acceptable to derive them from donated IVF-embryos or embryos established through therapeutic cloning. In relation to IVF-embryos it argues that since they are normally 'surplus' to requirements, and will not be implanted in a uterus, there are no moral objections to the use of such embryos to create a stem cell line, provided that parental consent is obtained to this further use. The alternative is to allow the embryo to perish.

The Council further argues that granted that the embryo (IVF or cloned) has no future life and does not benefit from any of the research, it loses nothing, whereas other future persons might benefit from therapeutic research using ES cell lines (Nuffield Council 2000, paragraph 22).

Activity:

(a) Do you believe that using spare embryos for research or creating them for this purpose makes a difference from an ethical point of view? Make a list of the ethical differences (if any) that you can identify between sourcing stem cells in spare IVF-embryos or specially creating embryos.

(b) A central ethical objection to therapeutic cloning is that human embryos are created solely for instrumental use. Whether or not this can be justified – and under what conditions – is an ongoing issue of debate. In focus then, the main questions of this chapter are: Is it ethical to allow embryo research? Under what conditions? Should only spare embryos be used or should embryos be created and destroyed for this purpose?

(c) Even if you do not accept germ line therapy and/or therapeutic cloning, do you think that you could not accept ES cells research because this may ultimately lead to them? Can we just prohibit research because of fears about possible futures? Should these fears be well founded? It is often difficult to distinguish our fears stemming from rapidity of developments and fears premised on empirical probabilities.

So far, we have examined the slippery slope argument in relation to the moral acceptability of ES cell research on surplus IVF-embryos. Now, we will evaluate the 'other options' argument which de Wert and Berghmans consider has an important application to the ethical debate on this issue. They define the 'other options' position in the following way.

The 'other options' argument'

A condition that might justify the instrumental use of embryos is that no suitable alternatives exist that may serve the goal(s) of the research. However, critics of this kind of justification claim that at least three alternatives exist which do not require the instrumental use of embryos:

1. **xenotransplantation**, involving the transplantation of tissues and bodily parts from nonhuman animals to humans
2. the use of ES cells from another source, i.e., human embryonic **germ cells** that may be harvested from tissue after elective abortion; and
3. the use of so-called 'adult' stem cells

The question is not whether these possible alternatives require further research (this is uncontestable), but whether only these alternatives should be researched. In other words, keep the following question in mind: Since we can explore other alternatives and since instrumental use of embryos is ethically problematic, is a moratorium for research into preimplantation embryo ES cells called for? Or, is it preferable to research the different options in parallel with preimplantation embryo research?

1. Xenotransplantation

Some critics argue that xenotransplantation is still an experimental technique and implies many risks to human health. The US Committee on Xenograft Transplantation: Ethical Issues and Public Policy (Institute of Medicine 1996: 1) states that 'the scientific advances and promise, however, raise complex questions that must be addressed by researchers, physicians and surgeons, health care providers, policymakers, patients and their families, public health officials, the news media, and the public'. The Committee recommends further investigation into a number of ethical issues that are raised by the procedure of xenotransplantation such as the standard of informed consent that must be secured (since lifetime surveillance of patients will be needed), the issue of fairness and justice in allocating organs, as well as the psychological and social impact of receiving animal organs on recipients, their families, and members of society as a whole.

2. The use of embryonic germ cells

Dolores Dooley (2000) highlights a number of concerns that some people might have with the use of embryonic germ cells.

Since aborted foetuses will undoubtedly prove to be a major source of ES cells we face questions about the moral connections between the activity of deriving stem cells or therapeutically using stem cells from dead foetuses. Does the derivation and use of stem cells from aborted foetuses implicate agents in the act of abortion? Does using foetuses as a source of ES cells weaken moral objections some may have to abortion by seeming to achieve a good end from what some deem the morally offensive act of abortion? Put otherwise, if one objects to abortion or otherwise thinks we should not encourage abortions, should we be concerned that researchers and patients benefiting from foetal stem cells are accomplices in abortion?

Notice that the process of procuring primordial germ cells from aborted foetuses does not kill the foetus. So what would be the act of the accomplice in this objection to use of abortus stem cells? Deriving the stem cells is done after the death of the foetus but while cells are still viable. However, even if the accomplice challenge is dissipated, opponents of abortion might have a residual concern that foetal cell derivation just might make abortion appear to be a 'positive act'.

Activity:

What do you think of the 'accomplice' argument here? Why can we not separate the abortion decision of a woman from the deriving of stem cells from the aborted foetus, and primarily aim to help others by therapeutic stem cell use without influencing or encouraging abortion? Do you think this results in the encouragement of abortion?

Notice that this is another variation on a 'slippery slope' position: the ethical concern that the therapeutic use of aborted stem cells might lead to encouragement of abortion. If one objects to abortion, then one should also object to the use of abortus stem cells.

On the same issue Sigrid Graumann (2000a) speaks to the question of political conflict which arises if the use of ES cells touches on deeply held moral convictions. Graumann claims below that conflict is inevitable and unavoidable.

Thus, the main problem seems to be the fact that we are either using spare embryos *in vitro* to derive ES cells or aborted foetuses to derive EG cells. We thus become embroiled in two of the most controversial topics in ethical debates:
(a) the question of embryo research and
(b) the question of abortion

It has not been possible to resolve these controversies in many years of ethical debate, and I really don't believe that we will be able to solve these problems in the ethical discussion of stem cell research either. I think we should just face the problem that by allowing ES cell or EG cells research in one way or the other, the deep moral convictions of certain people will be violated. Our aim should be to find ways to deal with the political conflicts that result.

Activity:

What do you think Graumann is suggesting here? How might a society deal with the political conflicts arising here or in relation to other ethically sensitive issues?

Finally, Edmund Pellegrino (1983: 163) gives the Roman Catholic view on the subject which expresses a non-instrumental position that does not allow for stem cell research even though the benefits are admitted.

> Although there is a difference in moral gravity in harvesting cells from aborted foetuses if the act of ending a life is clearly separated from the use of the cells, the moral problem remains because Roman Catholics believe abortion to be intrinsically wrong.

Activity:

(a) Does Pellegrino deal with the slippery slope implied in his text here? Even if, as Roman Catholics hold, abortion is intrinsically wrong, it is still very controversial to claim that encouragement of abortion would result from using stem cells from the aborted foetus? Would you agree with Pellegrino's views? If so, how would you answer the challenge just put to it?

(b) Granted that some may argue that the use of stem cells from IVF surplus embryos is a much more direct and intentional assault on the human embryo (since they are destroyed in the process) the discussion above concerns the ethical acceptability of deriving stem cells from aborted foetuses. In abortion, the foetal embryo is destroyed by the intentional decision of the pregnant woman and the abortus with all cellular materials will be incinerated if not used. Do you think that one can consistently object to abortion but utilise aborted stem cells? Give reasons for your answers.

3. Adult stem cells

The US National Institutes of Health argue that although adult stem cells give real promise, their use does not present ethical problems and does not involve the destruction of human life, there are limitations to what we may or may not expect from them. Firstly, there is no evidence that adult stem cells have the broad potentials of pluripotent stem cells and secondly, these cells are often present in only minute quantities, are difficult to isolate and their numbers may decrease with age (US National Institutes of Health 2000). However, it is important to note that, at this time, it is the adult human stem cells that are sufficiently well-enough understood that they can be reliably differentiated into specific tissue types and proceeded to clinical trials.

Comparing the options for sourcing ES cells

So far, we have examined xenotransplanation, aborted foetuses and adults as possible sources of stem cells. Let's continue now with de Wert's and Berghmans' discussion of these options and the reasons for their claim that a moratorium on embryonic stem cell research should not proceed.

For a comparative ethical analysis of ES cells of preimplantation embryos on the one hand and the above discussed alternatives on the other hand, a number of relevant aspects can be distinguished:

- the burdens and/or risks of the different options for the patient and his or her environment
- the chance that the alternative options have the same (probably broad) applicability as ES cells from preimplantation embryos; and
- the time-scale in which clinically useful applications are to be expected

1. *Xenotransplantation* entails the risk of cross-species infections and the resulting threat to public health. This risk, at least for the time being, is an ethical threshold for clinical trials. Above that, from a perspective of animal ethics, the question may be raised whether it is reasonable to breed and kill animals in order to have transplants, when at the same time human spare embryos are existent which otherwise would be discarded.

2. The use of *embryonic germ cells* from a moral perspective seems – all other things being equal – to be more desirable than the instrumental use of (living) preimplantation embryos. For the time being, however, EG cells are (very) difficult to cultivate. Above that, animal research suggests a disturbed reprogramming of these cells, which makes the clinical application of these cells less desirable because of possible health risks for the recipient.

3. Concerning *adult stem cells*, the most optimistic expectation is that only in the long run may these appear to have an equal plasticity as ES cells (and be as broadly applicable in the clinic).

If ES cells from preimplantation embryos in the short term will have more potential clinical applications, then the risk of a moratorium is that patients are potentially harmed. This in itself is one reason not to proceed with a moratorium. The simultaneous development of different research strategies is preferable, considering that research on ES cells will probably contribute to speeding up and optimising clinical applications of adult stem cells. Maybe in the future it will appear that at least some (types of) adult stem cells (e.g., in bone marrow) are, in fact, 'lost' ES cells. In that case an adult patient might be an alternative source of ES cells. This source certainly would be preferable to embryo use, presuming that (also these) ES cells are not morally equivalent to an embryo, and, on the condition that these cells could be harvested in a way that is not seriously burdensome or risky to the patient.

Activity:

To recap, the focus of this paper is on the debate that surrounds the sourcing of ES cells: surplus embryos from IVF? created embryos (from IVF or cloning)? aborted foetuses? adults or animals? So far, we have examined arguments in relation to ES research on surplus IVF embryos that concern:

• the slippery slope argument and
• the other options argument

Having now read the de Wert and Berghmans discussion on embryo research, are you able to come to a conclusion about the issues developed in this chapter? Should embryo research be allowed? Should we go on with ES cells research? Are there good reasons for a moratorium on embryo research for stem cell procurement? Give reasons for any of your answers.

The position that de Wert and Berghmans elaborated, that is, that we should simultaneously pursue all lines of stem cell research and search for the best sources of them – is a view shared by many. But, there are other dissenting views. Gilbert Meilander (2001) lists the available options in the following way:

> We might be convinced that the goal of relieving suffering offers a straightforward justification of sacrificing embryos in research. Or we might hold that although embryos merit our respect, the greater the good to be achieved by destroying them, the more respect must give way to research. Or, if we take the notion of respect seriously, we might find that relieving suffering is a real but not supreme imperative. (Meilander 2001: 9)

In this paragraph Meilander argues that proponents of embryo research [such as the Nuffield Council or de Wert and Berghmans above] justify it in utilitarian terms, on the basis of the goal that can be achieved and the relative worth of the embryo. So if we clarify Meilander's points, we see that the first utilitarian line of defence prioritises the goal of research, the second [which we can attribute to de Wert and Berghmans] takes a more moderate approach because it gives some (relative) weight to the embryo. In both cases, the therapeutically expected goal of research justifies the means used in research. Meilander's preferred option is the third position: he accepts that the goal of relieving suffering is an important one, but claims that this therapeutic goal does not outweigh the value of the embryo. For Meilander, the embryo has intrinsic value, it cannot be used as a means involving its own destruction for any end no matter how valuable that end might be.

Study groups on embryo research have also recommended prudence and caution in relation to research on therapeutic cloning. For example, as recently as November 2000 a report by the twelve-member European Group on Ethics in Science and New Technologies (EGE), which answers directly to the President of the European Commission, accepts the major research interest in human stem cells and their promising therapeutic prospects, but nonetheless considers the creation of embryos by SCNT premature. The Committee advises a cautionary approach to any legislation that would permit the creation of embryos for research purposes and, in particular, for the purposes of SCNT. In relation to SCNT the group concluded that

> these remote therapeutic perspectives must be balanced against considerations related to the risks of trivializing the use of embryos and exerting pressure on women, as sources of **oocytes**, and so increasing the possibility of their instrumentalisation. (EGE 2000: 17)

The prudence expressed by the EGE reflects the diversity of opinion among its members in relation to the status of the embryo and the instrumental use to which it can justifiably be put. (See Appendix 2 for a summary of the legal situation in the Member States.)

Similarly, the European Convention on Human Rights and Biomedicine prohibits the creation of human embryos for research purposes while some of the areas of research which the US National Institutes of Health (2000) considers ineligible for funding include:

- research in which human pluripotent stem cells are derived using SCNT
- research utilising human pluripotent stem cells that are derived from SCNT
- research in which human pluripotent stem cells are used in combination with SCNT for the purposes of **reproductive cloning** of a human

Most people would argue that even if we agree that ES cells research is ethically acceptable in principle, it should be regulated by certain conditions stipulated in law or professional guidelines. We suggest that these conditions should address the following concerns:

- the protection of the freedom of choice and health of women who donate embryos
- the status of the embryo
- the development of consent procedures and restrictions on the donation and sale of the tissue to prevent inducement to terminate pregnancies or dispose of embryos

Furthermore, while there are many strong and influential voices from the scientific and medical community urging us to proceed, there are also wider social concerns e.g.:

- concerns about social justice and fears that such great benefits will be distributed unevenly
- the need for informing and educating the public about the ethical and policy issues raised by stem cell research and its applications

Activity:

(a) Now that we have come to the end of this chapter think again about your views on embryo research. Which of Meilander's options comes closest to your own position? How important is the goal of the research? Does a good end justify interventions of the kind required in the discussion above?

(b) If you consider that embryo research is morally acceptable in some cases, are there any limits or conditions that you would place on that research?

To briefly summarise the positions already outlined, consider the following arguments for and against a moratorium on stem cell research involving human embryos. See if you can construct any more arguments of your own or, in fact, whether you would take issue with any of the arguments offered.

Arguments for and against imposing a moratorium on stem cell research involving embryos

For

✔ access to cell replacement therapy is not a sufficient good to justify the destruction of human embryos

Against

✗ stem cell research promises significant therapeutic benefits

✔ even if embryos are destroyed anyway in the IVF process, to use them for stem cell sourcing is to further instrumentalise them in an unacceptable way, treating them solely as means to some desired end but in no way benefiting them

✘ many of the embryos used for stem cell use would otherwise be destroyed. It is better to benefit than see no good from this destruction

✔ in pursuing embryo research for therapeutic ends we are mounting a slippery slope which would likely lead to reproductive cloning, a practice many would find ethically unacceptable

✘ the embryo merits respect but of a relative kind that may be over-ridden by concern for many future human lives to benefit from stem cell research

✔ the pursuit of other alternative sources for stem cells can be vigorously undertaken and thus we can benefit therapeutically in the long run while not requiring the use and destruction of embryos

✘ thus far, the alternative sources of stem cells do not promise the same success and therapeutic benefits. Further research on embryo stem cells will assist in understanding the other alternatives for stem cell sourcing

✔ ?

✘ ?

Activity:

Which position do you find most convincing and on what grounds?

Summary

This Chapter addressed the controversial topic of embryo research, specifically embryonic stem cell research. The Nash case can be reviewed before you finish because that case began this chapter and is also controversial, in part, because the destruction of thirteen embryos seemed necessary as a means to the end of getting an embryo (later called Adam) that would be effective as treatment for Molly Nash. In subsequent arguments you read claims that the overreaching goal of ES research is the relief of human suffering but that too is not an unqualified good. Rather it is ethically controversial because it involves the use and/or creation and/or destruction of human embryos. The ethical debate centered on whether or not the goals of embryo research are of sufficient merit that they justify the instrumental use of embryos that is required to achieve it.

Chapter 5 will address a separate but related issue: whether or not pregnant women should be considered to be the owners, in some sense, of the embryonic tissue (and in turn the stem cells that are derived from it) that they have laboured to produce.

– 5 –

Women and the Commercialisation of Embryonic Tissue

Objectives

At the end of this chapter you should be able to respond to the following questions:

- Is a stem cell line the kind of thing that can be owned?
- If it is considered property in some sense, is it the exclusive property of the scientists and institutions/corporations who develop it?
- Does the pregnant woman have a prior claim to a stem cell line because of the labour she has contributed to the creation of the embryonic tissue from which it is sourced?

Chapter 4 considered the moral acceptability, limits and conditions of embryonic stem cell research from a 'foetalist' perspective (focusing on the moral status of the embryo or foetus from which stem cells are derived). This Chapter is centrally about the interests of women with regard to embryonic stem cell lines and not just on the origin of stem cells *per se*. The Chapter will first of all involve an exercise around extracts from a paper by Donna Dickenson who argues that (a) stem cells are a kind of property and (b) pregnant women have some property rights to them because of the labour they put into producing them.

Before starting with Dickenson's analysis let's look at a very concrete example of what is at stake here.

The Case of Sharon

Sharon wants to conceive using IVF even though she is not suffering from infertility difficulties that might therapeutically warrant IVF. In undergoing IVF she is eager to harvest stem cells from two to three healthy surplus embryos for later therapeutic use – to benefit her own health or the health of her future child. She comes to you for advice.

Activity:

Does Sharon have this entitlement? As you will see, using Dickenson's argument, one might think so. What do you think?

Now read the first part of Dickenson's paper.

Who owns foetal tissue?

Donna Dickenson

Until very recently the question of who owns embryonic or foetal tissue was of limited commercial importance, although there were applications of aborted foetal tissue in the treatment of Parkinson's disease, diabetes mellitus and other conditions. With a few exceptions the use of embryonic tissue was, so to speak, a non-issue.

By mid-1999, however, commercial exploitation of stem cells had been termed the most controversial ethical issue in biomedicine (Capron 1999). The threshold event occurred in November 1998, when two separate teams of scientists in the United States (US) claimed that they had managed to isolate pluripotent human embryonic and foetal cells and grow them indefinitely under laboratory conditions. Because of the extraordinary therapeutic promise that stem cell research holds and also because the use of stem cells would streamline pharmaceutical testing (new drugs could be tested for safety and efficacy on cultured stem cells before being tested in humans); pharmaceutical and biotechnology firms are hugely interested in the use of such cultured cells and in the development of tissue banks of both undifferentiated and specialised cells and tissues. Six to eight pluripotent cell lines have already been developed, in the US and Singapore: none at present in the United Kingdom (UK). Eventually the need for embryo 'donations' should lessen as the self-replicating stem-cell lines grow in size. But does this mean that the ethical issues will disappear? Hardly: the enormous commercial value of these cell lines, which will increase with their size, raises profound issues of justice and exploitation, particularly issues of property rights.

Both the US teams were funded by the Geron Corporation, a US biotechnology firm which is now seeking a patent on the technologies. Meanwhile, the UK's Roslin Institute, which produced the Dolly cloning technique, was reported to be exploring collaboration with the Wisconsin researchers, with a view towards deriving cells from adult patients that could be cloned using isolated embryonic stem cells. The aim is to develop cell therapy rather than manufacturing tissues and organs, with the advantage of avoiding immunological rejection problems. By a remarkable coincidence, the Geron Corporation also has a major interest in the commercial arm of the Roslin Institute. The globalisation of stem cell research and application is already upon us. It will almost certainly mushroom into an international trade in embryonic stem cells: already German research groups are using embryonic stem cells imported from other 'less moral' countries such as Denmark, Finland, Spain, Sweden and the UK, since the German Embryo Protection Act of 1990 prohibits any retrieval of cells from embryos, under criminal penalties (Lunshof 2000).

One of the US teams, James Thomson's group, based at the University of Wisconsin, used embryos grown *in vitro*, developed through fertilisation of the mother's ova with the father's sperm: primarily 'spare' embryos which are not to be implanted. Since embryonic stem (ES) cells are derived from the inner cell mass of the blastocysts and blastocysts are used before implantation, the mother's 'sweat equity' is reduced, but she has still undergone the labour of stimulation with fertility drugs (superovulation) and extraction of ova – painful and moderately risky procedures.

John Gearhart's team at Johns Hopkins used embryonic germ (EG) cells, derived from the reproductive cells of embryos aborted five to nine weeks after fertilisation. This method of using EG cells relies on aborted foetal tissue, into which the mother has put the labour of early pregnancy.

Interest has mainly focused on ES cells which have the capacity to be maintained indefinitely and the potential to develop into every cell of the body rather than EG cells, where attempts to derive adult cells in mice have led to abnormalities. Use of ES cells has also been assumed to be less ethically debatable, because it does not require abortion, or because pre-embryos (under fourteen days of development) are thought to have a lesser moral status than embryos or foetuses. I think it is far from obvious, however, that use of ES and EG cells is ethically trouble-free; but that is because I shall focus, not on the status of the embryo/foetus, but on the rights of the mother.

Discussion

Dickenson's paper focuses on the method of producing ES and EG cells and the rights of the pregnant woman. Specifically, she does not engage with any of the questions we have already discussed in Chapter 4 that arise in relation to the status of the embryo which is the source of these cells. However, if we allow that embryo research, and by implication, stem cell research, is not ruled out entirely on the basis of appeal to the moral status of the embryo, we can continue with Dickenson's argument and consider a further reason why we should be cautious in relation to this issue. In her paper so far, she has drawn attention to the globalisation of stem cell research through, for example, the activities of the US based Geron Corporation and the international trade in stem cells that she anticipates is likely. In the following extract she focuses on the labour that women put into the production of embryos and argues that this labour entitles them to a claim on the stem cells derived from either the aborted or 'surplus' embryos that they produce.

The remainder of this paper will concentrate on the risks of exploitation to pregnant women, and conversely on the arguments in favour of their possessing a property right in

stem cells derived from their embryos or foetuses, in addition
to the procedural right to give or withhold consent to the
further use of those tissues. This new focus is particularly
urgent if the UK does implement the Nuffield Council recom-
mendation to allow research involving human embryos for
purposes of developing tissue from embryonic stem cells,
amending existing legislation (Schedule 2 of the Human
Fertilisation and Embryology Act [HFE Act]). At present there
is no plan to create embryos deliberately for this purpose,
provided that sufficient such cells can be obtained from
donated surplus IVF-embryos. However, even this compara-
tively modest proposal raises difficulties about possible
exploitation of vulnerable couples undergoing IVF, and particularly about the interests of women.

Property, Persons, Pregnancy and Progeny

Even if we concede that the embryo or foetus has no rights
which could give rise to duties to refrain from stem cell development, that says nothing whatsoever about the rights of the
parents, and particularly about the rights of the woman.
These rights can be viewed after the fashion of John Locke, a
seventeenth-century philosopher, as derived from the labour
which women put into the processes of superovulation and
egg extraction (ES cells) or early pregnancy and abortion (EG
cells). Alternatively, a marxist-feminist interpretation would
emphasise the added value which women put into the 'raw
material' of gametes. If the marxist interpretation is preferred,
one would focus interest on women's alienation from their
reproductive labour, and on the exploitative transfer of rights
in the products of that labour to private commercial companies such as biotechnological firms or IVF clinics. For the
purposes of this discussion, either the liberal or the marxist
model is valid. However, most of my discussion will be more
Lockean than marxist.

It is important to emphasise that Locke finds the notion of
property in the body unsatisfactory: he does not say we own
our bodies; only God does. What we do own is our labour,
which is the expression of our moral agency or personhood;
and this is what Locke is referring to when he declares that
'Every man hath a property in his own person' (Waldron
1988, Dickenson 1997). The popular form or liberal argument

about ownership of the body rests on Locke's premise that we cannot own our bodies because they are the instrument by which we create other wealth, and which therefore predate and ground our ownership of the other things we create. To extrapolate this argument to organs and tissue: no other parts of the body are owned, because we do not put labour into creating our own bodies.

If the notion of property in the body is philosophically unsatisfactory, it is equally scorned by the common law tradition. Broadly speaking, common law views tissue taken from the body not as the property of the person from whose body it comes, but as *res nullius*, no one's property. What the law has traditionally been concerned with was making sure that the tissue was not taken without consent, not with what happened to it afterwards – after all, it was presumed to be diseased. Property law, by contrast, can be concerned with subsequent rights, immunities and other forms of control following the property transfer. One example of the emphasis that the law places on the notion of consent as opposed to property is illustrated by the decision taken in Moore v. Board of Regents of the University of California (1990). This decision related to the removal by doctors at the University of California in 1976 of the enlarged spleen of John Moore who was suffering from a rare form of cancer. Following on his splenectomy, researchers at the university developed a productive cell line, the 'Mo' cell line, from Moore's T-cells. What we have in the Court decision is an excellent example of the old maxim, 'Hard cases make bad law'. This case exhibited such egregious abuses of the patient's informed consent that the legal judgement turned almost entirely on those abuses. (Not only was Moore never told that his original splenectomy had yielded T-cells with remarkable immune powers, developed into a $3 billion cell line; he was asked to keep returning to donate all manner of other bodily products on the pretext of further therapy and check-ups.) The wider issue of development rights in Moore's cell lines was decided in a manner which arguably breaches previous legal precedents, by awarding all such rights to the researchers and the hospital board of regents (Gold 1996). It may be that the patient does not own the tissue, but does that necessarily mean that the researcher or hospital does? If we wish to avoid commodification of tissue by allowing the patient tradeable rights, why are we so willing to allow commodification by allowing tradeable rights to the researchers?

The presumption in the case of embryonic stem cells and embryonic germ cells, however, must be that the woman has a presumptive right in these cells, outweighing the rights of the clinic and researchers. If we grant, following Locke, that we only have a claim to that into which we have put our labour, and we accept that the pregnant woman puts labour into producing foetal tissue, I would argue, she can be said to own it in some sense. (See Dickenson 1997 for a fuller version of this argument.)

In the case of both ES and EG cells, of course, the father has also made a genetic contribution, but, I would argue, not a donation of his labour. It is obscene in more than one sense to compare masturbation to produce sperm with superovulation and egg extraction. To argue for the father's ownership of either blastocysts grown *in vitro* or embryos, then, one would have to assert that his right derives from his ownership of his sperm or genes, but I have already argued that we do not actually own parts of our body, including gametes and genes. The 1990 Human Fertilisation and Embryology Act (HFE Act) supports this interpretation insofar as it pays gamete donors expenses; the Authority and the clinics it licenses are not purchasing gametes, but recompensing donors for loss of time and travel expenses.

Activity:

The pages you have been reading have introduced some fairly dense philosophical argument into our consideration of how we should view the ownership of cell lines. Take a little time to consider the main points of Dickenson's argument. She claims that:

• The traditional emphasis of the common law on issues of consent rather than on issues of ownership is inadequate to deal with the possibilities for commercialisation of human tissue that have been opened up by stem cell research.

• Even though we do not actually own parts of our bodies, including stem cells, there is one exception. Women possess a property right in stem cells derived from their embryos in addition to the right to give or withhold consent regarding their further use.

- This property right appeals to a Lockean notion of property and is based on the fact that pregnant women put labour into the production of embryonic and foetal tissue.
- Even though fathers also make a genetic contribution to embryonic tissue, this contribution does not afford them rights of ownership over the tissue because it is not the outcome of their labour.

(a) What do you think of Dickenson's argument? Do you think that she is right to conclude that women's labour in pregnancy gives them property rights to the products of their labour?

(b) Some might argue that Dickenson oversimplifies and trivialises the contribution that men make to the production of blastocysts grown *in vitro*. What do you think? Even if we acknowledge that the woman's labour is greater than the man's, does this mean that, therefore, she has rights and he has none? Is it a question of degree? Or would you agree with Dickenson that it is 'obscene' to compare the labour of women and men in this case?

(c) Dickenson's paper goes on to acknowledge that current policies and practices favour a focus on informed consent rather than ownership. As you read, ask yourself which approach best serves the interests of all concerned. Can Dickenson's arguments be applied in your country?

Far from affording the pregnant woman rights in the embryonic and foetal tissue which she has laboured to create, most current policy documents concentrate on making sure that she freely gives up any such rights through giving a clear and separate consent to use of the tissue for research and therapeutic purposes. This exhortation in the name of the 'gift relationship' (Titmuss 1972) is the strategy suggested separately by the National Bioethics Advisory Commission Report (1999) in the US, and by the Medical Research Council working party and the Nuffield Council commission in the UK. The advisory report from the Geron Corporation ethics group is slightly more frank in advising that women donating embryonic or foetal tissue should be told about market value, but one suspects that this proviso is inserted merely to stave off Moore-type legal actions (Geron Ethics Advisory Board 1999).

Given the vulnerability of IVF patients, and their typical gratitude towards the clinicians for giving them any chance at a child, there is plenty of room for exploitation. As Lori Knowles puts its, there is a tension between the altruism individuals are supposed to exhibit by donating their tissue for research and the current patent system, which encourages companies to stake lucrative property claims in that research (Knowles 1999: 38).

Knowles goes on to claim that '[in] law, a tangible thing is either a person or property, and if it is one it cannot be the other' (Knowles 1999: 39). In the US and the UK it is widely agreed that the blastocyst is not a person. Therefore, following Knowles, it must be property, as the Geron Corporation is happy to agree; but it does not follow from this that the property necessarily belongs solely to the Geron Corporation or its equivalents. In the Moore case it was argued that granting any form of property rights to the tissue donor, Moore, would impede the free flow of scientific research – but so, of course, do patents on cell lines and genes by biotechnology companies. The issue of whose property, for whose benefit, needs a wider airing than it has so far received in the stem cells debate. As Knowles cogently states:

> Fears about a market in human body parts and about commodifying human reproduction have prompted many to suggest that couples should not sell their embryos. The same arguments are used to argue that donors should not share in the profits resulting from research on their embryos. In property law, however, restrictions on sales are prompted by the nature of the property itself, not by the status of the person claiming a commercial interest. Therefore, if it is wrong to commercialise embryos because of their nature, then it is wrong for everyone. It is simply inconsistent to argue that couples should act altruistically because commercialising embryos is wrong, while permitting corporations and scientists to profit financially from cells derived by destroying those embryos. (Knowles 1999: 40)

Activity:

(a) Take a few moments to re-read the quotations from Knowles. Consider the pressure that there is on women and couples to act altruistically with regard to their embryos. Do you think that a comparison can be drawn between this discussion and the debate outlined in Chapter 3 of this workbook? In that chapter, the debate centred on the social and legal conditions encouraging women who undertake surrogate pregnancies to do so for altruistic reasons alone.

(b) Note Knowles' claim that when property law places restriction on the sale of a certain kind of property, such restrictions are prompted by the nature of that property. Can you think of two examples of particular kinds of property whose sales are regulated by certain conditions? Examples might be the sale of cattle or endangered species.

(c) Do you agree with Knowles' claim that the status of the one who claims ownership of any given property is irrelevant? What do you think of her related claim that it is inconsistent to argue that couples should behave altruistically because of the nature of embryos, and, at the same time, permit the commercialisation of embryos by corporate scientists?

Now read the last section of Dickenson's paper which outlines and critically evaluates three different solutions to the current impasse.

What are the models to follow in the stem cell debate? We could adopt one of at least three possible approaches to ownership of embryonic and foetal tissue:

1. Status quo

In this model we would uphold the law's primary concentration on obtaining consent from the donor of the tissue, rather than conferring property rights on her. This is the basis of guidelines from the American College of Medical Genetics, which establish that patients must be asked for consent before research is done on tissue samples (American College of Medical Genetics 1995). It was also roughly the approach of the UK Polkinghorne Committee (1989), although the scope of profitability and commercial application of tissue has moved on enormously since then. Given, however, that 'informed consent is no part of English law' this is unlikely to provide satisfactory protection for women (Sidaway 1985). The very favourable public image of IVF is another problem: there is not going to be much pressure on IVF clinics to justify what they do with blastocysts obtained as part of infertility treatment. Couples may be pressured to agree that 'spare' blastocysts can be used for commercial purposes, perhaps in exchange for reduced cost of treatment cycles.

The first model continues to maintain the fiction that tissue extracted after procedures is no longer of any interest to anyone. Yet between 1976 and 1993 Merieux UK collected 360 tons of placental tissue annually from UK hospitals for sale to French drug companies (Nelkin and Andrews 1998). Almost certainly, none of the mothers was asked for her consent to this use of the placenta grown in her body, and expelled as the final stage of her labour in childbirth. In Canada, a similar practice was reversed after a Sicilian woman asked for the placenta, in order to carry out the custom of eating it; only then did the extent of the scandal become known (Baudouin 1999).

Similar issues arise in relation to umbilical cord tissue. In 1988 a French team under Dr Elaine Gluckman developed a process for turning umbilical cord blood into a substitute for bone marrow in transplantation. The team originally envisaged communal, nonprofit banks of umbilical cord blood, but the process was quickly taken over by private firms, who marketed their own reprocessed blood back to the mothers as a form of insurance, to be stored for their babies (Sugarman et al. 1997).

In the face of full-scale commercialisation elsewhere of life forms, following the 1980 US decision in Diamond v. Chakrabatty and the 1998 decision by the European Parliament to support patenting of life in order to maintain competitiveness with the US, we need better protection than the common law has previously afforded us. The amount of original input necessary to obtain a patent is minimal: for example, a patent on a diagnostic test for Down's syndrome was given to a researcher who merely established a correlation between a particular hormone level and the syndrome – not the test itself. Researchers have been given patents on particular gene sequences without even having established their function. Not much labour has been 'mixed' with the natural substance in these cases. This leads us on to option 2.

Before moving on to option 2, take a few moments to consider the reasons why Dickenson is dissatisfied with the status quo model. She argues that because the model is concerned with obtaining consent from the donor of the tissue, rather than conferring property rights on her, it is not sufficiently adequate to protect the interests of women. This is because:

- there may not be any explicit doctrine of informed consent in the jurisdiction in which a woman lives (such as the UK);
- the favourable public image of IVF and of IVF clinics means that there is little pressure on them to justify what they do with spare embryos in comparison with the pressure on couples to part with embryos perhaps for reduced cost of treatment;
- the model obfuscates the fact that embryonic tissue is of significant scientific and commercial importance;
- in many instances patents to different forms of life such as gene sequences are being granted to researchers who have contributed very little 'labour' to the natural phenomena that they are dealing with.

Are you persuaded by Dickenson's reasons to reject the status quo model? Can you think of any additional reasons why it should be rejected? Can you think of any arguments in favour of accepting it?

Now go on to read the second option that Dickenson discusses.

2. Strict regulation of commercialisation

So far this second approach has not manifested itself much in practice in the US or UK, and it is likely to face even greater obstacles as commercial interests gather further momentum. To some considerable extent, however, the UK already regulates assisted reproduction, through the statutory regulatory body established by the HFE Act 1990, the Human Fertilisation and Embryology Authority (HFEA). In vitro research on human embryos is illegal without a licence from the Authority, for both the project and the premises in which it operates. The uses of foetal tissue (relevant to EG cells) are regulated by guidelines set down by the Polkinghorne Committee

(1989), aimed at maintaining a strict divide between the decision to undergo an abortion and the decision to allow further uses of aborted tissue. The Polkinghorne Review (1989) also concluded that research ethics committee approval must be sought for 'all proposals for work with foetuses or foetal tissue, whether alive or dead, and whether classed as research or therapy, because of the high level of public concern.' In the wake of recent scandals, such as the Bristol and Alder Hey hospital enquiries over the retention of dead children's organs, the public is no doubt still concerned. But there are no statutory provisions in the UK governing the uses which can be made of aborted foetal tissue; as with the Moore judgement, the focus is solely on the correctness of the procedure, not on the uses made afterwards of the 'tissue' removed.

No research can be authorised on embryos older than fourteen days, but that provision would still allow the method developed by the Wisconsin team, using blastocysts, under license from the HFEA. Nonetheless, there are mounting commercial pressures, by which I include pressures from the leading IVF clinics, to repeal the fourteen-day rule and to end regulation altogether. Moreover, the HFE Act is flexible enough to permit use of embryonic stem cells, proponents of commercialisation are already arguing – even though the techniques described in this paper, and the cloning technologies which give them commercial importance, were completely unknown at the time the Act was passed in 1990. The arguments for therapeutic benefit of stem cell commercialisation seem to be in the ascendant in the UK at the moment, which does not augur particularly well for strict regulation.

Some commentators (e.g., Gold 1996, Knowles 1999) have proposed that commercial researchers and firms should be permitted to 'commodify' stem cells and other bodily tissues, but only under the condition that they return a share of the profits to the National Health Service or the wider political community. In this model, donation would remain altruistic, but firms would be obliged to make cell lines widely available and to price the products derived from them at an affordable level – under pain of penalties from a patented biotechnology products review board. As a pragmatic solution, this proposal is attractive, but I want to propose something different.

Activity:

(a) Before we continue with Dickenson's proposal, take a few moments to consider your response to her worries that any regulation of stem cell research will be difficult to achieve and that any of the limits and conditions set by regulatory bodies are being continually eroded. Dickenson focuses on the push-pull relationship between the regulatory bodies and commercial interests in the UK, but can you think of any comparable developments in your own country?

(b) What is your response to the utilitarian argument against decommercialisation that much good which could be done if the profit motive is allowed will not be done if firms are barred from owning, or strictly controlled in their use of, life-forms?

Now return to evaluate Dickenson's alternative.

3. Vesting control over all tissue in the mother, and treating alienation of it from her as theft

A feminist model, sensitive to women's alienation from their reproductive labour, might want to take a more radical tack than regulating commercial interests. A recent decision in the UK case of R v Kelly might be construed favourably to women in this regard. In the Kelly case the Court of Appeal ruled that an artist who had used parts of dead bodies had committed a criminal offence against the Royal College of Surgeons, which had preserved them. Although the Court reiterated that there is no property in a dead body as such, it acknowledged that the limbs had been transformed into artefacts which could be owned, by application of the labour and skill of dissection and preservation for teaching purposes – a good Lockean argument: 'Parts of a corpse are capable of being property within s4 of the Theft Act if they have acquired attributes by virtue of the application of skill, such as dissection or preservation techniques, for exhibition or teaching purposes' (R v Kelly 1998).

The important thing to note here, less the grisliness of the case, is the way in which property is not construed as an all-or-nothing concept. To come under the Theft Act is sufficient for these purposes for a bodily part to be protected as a form of property (Grubb 1998). I suggest that it may also be sufficient in relation to property in foetal tissue and other products of pregnancy, provided we make two assumptions:

1. That embryonic or foetal tissue is akin to parts of a corpse (even though it has never been a living person) and that putting labour into bodily parts of a non-living body conveys some sort of property right.
2. That the labour which the woman puts into superovulation and egg harvesting (in the case of blastocysts and other forms of human embryo) and into pregnancy and childbirth (in the case of umbilical cord blood and placental tissue) gives her a right over the tissue.

This need not be a full-fledged property claim, and given the legal and philosophical incoherence of the concept of property in the body, it probably should not be. It need only be as great as the scope delineated in Kelly: enough to protect the tissue from being appropriated by others, under penalty of the Theft Act.

Even with these two assumptions, there are still problems. Vesting control over tissue in the mother may not be sufficient to protect the woman from exploitation by commercial interests. Those interests surpass any in surrogacy, where, as we saw in Chapter 3 of this workbook, it is difficult enough to distinguish between allowing women to contract as equals and opening them to exploitation. Arguably, thinking of the mother as having any kind of property interest in foetal tissue or the tissue by-products of pregnancy is also false to the uniqueness of the relationship between the woman and the developing foetus (Mahowald 1994).

Nonetheless, the great advantage of this model is that it recognises what women do and endure in infertility treatment, pregnancy and childbirth. It gives them a property in the labour of their persons and the products of that labour. This is not the same as owning a baby, which is not what we are talking about in the case of embryonic stem and germ cells. It is difficult to believe that placental tissue could have been 'harvested' without anyone's noticing that the mother might have something to say about it. Yet the ignoring of women's labour is pervasive throughout the discussion of rights over foetal and embryonic tissue (Mahowald 1994). The third model makes sure that women's labour gets notice.

If the second model is unlikely to succeed, however, why should the third have any chance at all? One reason is that it foregrounds the need to assign an owner to the tissue, that is, the bankruptcy of the traditional doctrine of *res nullius* in the face of commercial interests which want to make very sure that the *res* is definitely not *nullius*. Whereas regulation, in model 2, accepts that commercial interests or academic researchers own the tissue, but must bow to a certain degree of societal control over their actions, the more radical model actually affords a better chance of litigation establishing that their ownership is not free and clear. If all the women whose placentas were 'harvested' had to be compensated or indeed acknowledged, that would be quite a disincentive.

Near where I live, a motorway was planned to cut through an area where rare butterflies abounded, 'Alice's Meadow'. Hundreds of local people each bought a one-metre square of the meadow, and all their claims had to be adjudicated before eminent domain could be given. The motorway went elsewhere. It's a thought, isn't it?

Activity:

(a) Dickenson makes a strong argument in favour of the idea that women have a prior claim to the stem cells that are derived from the products of their labour. Do you find her argument convincing? Stem cell researchers might respond that while a woman's labour contributes to the creation of embryonic tissue it is the scientist's labour to develop the stem cell lines which, in turn, makes the embryonic tissue valuable. On this account, they would conclude that their claims and interests should be privileged over those of pregnant women. How would you respond to this argument?

(b) Now that you have read Dickenson's paper, look again at Sharon's case. Do you think that Dickenson would consider that Sharon is entitled to conceive by IVF in order to harvest stem cells for future therapeutic use? Dickenson's thesis is that women own their embryonic tissue in some sense. However, she argues that women do not have a ' full-fledged property claim' to the tissue, that their claim is 'only enough to protect the tissue from being appropriated by others, under penalty of the Theft Act'. Here, Dickenson argues that women's claim to embryonic tissue is limited and conditional: they are entitled to an interest in the tissue only if others make a claim to it. All else being equal, Dickenson might agree that there should be some conditions on the way in which women treat embryonic tissue. What kind of conditions (if any) do you think should be put in place?

Property and consent

As Dickenson's paper acknowledges, the philosophical and common law traditions have rejected the idea that we can own our bodies; they view the body as *res nullius*, no one's property. Instead, the focus of morality and law in relation to the removal of bodily parts and tissues has been on the requirement of informed consent. The requirement of consent is intended as a measure to protect the autonomy and privacy of the individual; one's right to control over one's body and medical treatment:

> What the law has traditionally been concerned with was making sure that the tissue was not taken without consent, not with what happened to it afterwards – after all, it was presumed to be diseased. Property law, by contrast, can be concerned with subsequent rights, immunities and other forms of control following the property transfer. (Dickenson, p.129 above)

Dickenson argues that the emphasis of the common law on issues of consent rather than on issues of ownership is inadequate to deal with the possibilities for commercialisation of human tissue that have been opened up by stem cell research. She cites the judgement in the Moore Case as one which set a bad precedent for subsequent decisions because it focused on the fact that Moore's consent was not secured rather than on his possible rights to the cell lines his tissue produced (these were awarded to the researchers and hospital board involved).

The following case study, which concerns the cryopreservation of the ovaries of a dead woman, raises two of the same ethical issues that are highlighted by Dickenson in relation to stem cells: property and consent. The case is presented from the perspective of the administrator/regulator of the HFEA, but it should be clear how different people might be involved in the ethical discussion and the decision-making.

The Case of the Cryopreservation of Ovarian Tissue

It is Friday afternoon, 4.30pm, when the telephone rings. It is a doctor at a private licensed ART clinic. He asks whether he can freeze tissue from the ovaries of a 32-year old woman. I explain that this is a new procedure and a new question for the Authority. There are issues of safety and efficacy to consider, of how it fits in the regulatory framework, and what kind of consents would be needed, etc. It is a complex question. Can I get back to him next week?

'No,' he says, 'I need an answer inside about 20 minutes.'

'20 minutes? That's impossible. Why?'

'Because the lady died an hour ago. Her husband is sitting in front of me and her ovaries are in a jar on my desk. I must freeze them in the next 20 minutes.'

I ask him to tell me some more about the background.

'The woman died from a very exceptional reaction to general anaesthetic during routine surgery. She was otherwise healthy. She did carry a multi-organ donor card with no exclusions specified. Her husband tells me that they have been trying for several months to achieve a pregnancy.'

'You had no consent to remove the ovaries?'

'No.'

'You have no consent to store them?'

'No.'

'You have no consent for her eggs to be used by another person?'

'No. '

'The husband plans to have her eggs fertilised and implanted in a surrogate?'

'Yes.'

'Technically, can you do that?'

'We don't know yet.'

(Whittall 2000)

Activity:

Consider your response to this case study if you were:

• the administrator/regulator responding to the doctor in the IVF clinic;
• the doctor making the request;
• the husband of the dead woman;
• the dead woman's sister or brother.

Now read Deirdre Madden's commentary on the case.

Commentary on the case of cryopreservation

Deirdre Madden

How can such a decision be made on the basis of one telephone call within a twenty minute timeframe? It is unreasonable to expect one individual to make a decision of such ethical importance without conferring with other colleagues and seeking professional and legal advice. How is such a decision to be made and by whom? Although legislative provisions and codes of practice may exist there will inevitably arise situations which are not explicitly covered by such provisions. Novel circumstances demand detailed analysis and discussion before a determination can be made. In many cases there will be sufficient time to allow consideration of professional and legal advice as well as consultation with an ethics committee. In cases where there is no time for such consultation, the decision-maker must use his/her own knowledge of the area to make a 'holding' decision, if possible, which should not commit him/her to any particular course of action in the future.

Is ovarian tissue to be considered property in some sense? Due to the strictures of public policy in, for example, the prevention of trafficking in body parts, what is meant by property here is something less than the absolute interests which one might have in ownership of other assets. The courts have not generally been in favour of the creation of property rights in relation to the human body, preferring to approach the issue from the angle of procreational autonomy or privacy.

What are the legal consequences of taking and freezing ovarian tissue without consent? In the Diane Blood case (*R v Human Fertilisation and Embryology Authority, ex parte Blood* [1997]) the taking of sperm from an unconscious man was considered to be a criminal offence. However the liability in criminal law was not explored as the sperm had been removed and stored in good faith and in consultation with the HFEA. It was also stated by the court that this was a situation that was not to be repeated. It may be argued that consent in this case should be presumed from the fact that the couple were trying to have a family of their own, (again as in the Blood case.) This evidence was found to be compelling in the Blood case even without any corroboration. It is however impossible to know whether the woman in this case would wish her eggs to be used for implantation in a surrogate mother. In the absence of such a stated intention, this procedure should not be permitted.

The use of a surrogate who will be the legal mother raises additional issues in relation to surrogacy and parentage due to the unique circumstances here. Is this an appropriate case in which surrogacy may be used? It is more usual for the child to be 'commissioned' by a couple, rather than a widower using his dead wife's tissue.

What of the best interests of the child in posthumous conception? Would it be a serious disadvantage to a child to be born in these circumstances with the combination of surrogacy and use of her dead mother's tissue? Although one may say that it is always better for a child to be born than not to be born, this confuses the situation with that in which it may be queried whether existing lives are worth living due to serious disabilities. The decision not to procreate does not harm anyone and therefore does not require a justification. However, a responsible decision to procreate requires careful consideration of the interests of the child to be conceived. (Madden 2000c)

Activity:

(a) Madden highlights some key ethical issues that arise in this case study, that:

- the decision in this case, like many ethical decisions, requires time for reflection and consultation with colleagues, professional and legal experts
- in some circumstances a 'holding' decision may be ethically appropriate
- the person best suited to make the ethical decision must be identified
- a property or consent approach must be taken to ovarian tissue
- the legal consequences of taking and freezing ovarian tissue without consent must be determined
- the appropriateness of using surrogacy in this situation must be weighed
- the best interests of any future child must be considered

She notes that it might be argued that consent in this case can be presumed because of the fact that the couple were already trying to have a family of their own. Would you agree that consent should be assumed? Do you think 'wanting a family' is grounds for assuming consent to the 'family' that the husband envisages?

(b) In this case the deceased woman had a multi-organ donor card with no exclusions specified. Could such a card be taken as granting consent for the removal (and donation) of her ovarian tissue?

(c) Assume that the ovarian tissue at the center of this case is not *res nullius* (and therefore governed by the requirements of informed consent) but a form of property. How might this change the outcome of the case? Given what you have read of Dickenson's position, do you think that she would accept ovarian tissue as a form of property in the Lockean sense?

In what follows we have compiled a table of some of the arguments for and against Dickenson's position that have been presented in this chapter. Take a few minutes to read the table and see whether you would take issue with any of the arguments offered. Can you construct any additional arguments of your own?

Arguments for and against the claim that pregnant women have property rights to embryonic stem cells

For	Against
✔ on a neo-Lockean view, embryonic stem cells can be considered to be a kind of property because labour is required to produce them. In addition, UK common law has recently established a precedent whereby some bodily parts have been treated as a form of property and, thereby protected under certain laws (R v Kelly 1998)	✘ the philosophical and common law traditions have rejected the idea that we can own our own bodies. These traditions would define an embryonic stem cell as *res nullius* and, therefore, as not the kind of thing that can be owned
✔ pregnant women have property rights to stem cells because of the labour they put into producing them – super-ovulation and egg extraction or early labour, abortion and childbirth	✘ thinking of the pregnant woman as having a property interest in foetal tissue is false to the uniqueness of the relationship between the woman and the developing foetus
✔ giving a property right to pregnant women in addition to their right to give or withhold consent to the further use of their embryonic tissue recognises what women do and endure in reproduction and guards against the exploitation of vulnerable women and couples undergoing IVF	✘ reproduction should not be commodified: the current policies, guidelines and practices of many countries, which focus on securing the informed consent of women to the use of embryonic tissue, protects women's autonomy and privacy and reflects the altruistic idea that tissue ought to be donated for research which benefits everyone

✔ the traditional doctrine of *res nullius* is bankrupt in the face of commercial interest in stem cells. Vesting rights to stem cells in pregnant women is a reasonable way to address problems that may arise regarding the commercialisation of human tissue and the globalisation of stem cell research

✘ even if we were to accept that stem cells can be owned, scientists and biotechnological firms have a greater claim to them because, while women may contribute the initial stem cells, these researchers and corporations develop the stem cell lines which give embryonic tissue the commercial value that it has

✔ in many instances, patents to different forms of life are granted to researchers who have contributed very little labour to the tissues that they are dealing with

✘ if commercial firms are not allowed to control and profit from stem cell research, they will not be motivated to carry it out and so, society would lose out on the therapeutic benefits of such research

✔ there may not be any explicit doctrine of informed consent in the jurisdiction in which a woman lives and so, attributing a property right to her may protect her interests

✘ men also make a contribution to the production of stem cells which give them a claim to any benefits that may follow on stem cell research

✔ ?

✘ ?

Activity

Which position do you find most convincing and on what grounds?

Summary

This chapter focused on two central questions in relation to stem cell research: (1) Can stem cells be owned? and (2) Do women have some kind of property right to their embryonic stem cells and a subsequent claim on any of the benefits that follow from stem cell research?

In the course of answering these questions we have
- examined the way in which notions of 'property' and 'consent' are deployed in the common law tradition
- discussed three possible approaches to the ownership of embryonic and foetal tissue and their implications
- applied what we have learned to a related issue: the removal and freezing of ovarian tissue for procreative purposes

– 6 –

Role Conflict
The Doctor as Treating
Physician and Researcher

Objectives
At the end of this chapter you should be able to:

- give reasons why mature human eggs are important for improving IVF treatment
- list some of the situations in which mature eggs might be harvested
- weigh the costs and benefits of such procedures
- critically discuss three arguments in favour of doctors asking some of their women patients to become egg donors

In Chapter 5 we addressed the interests of women in relation to the commercial development of stem cell lines for therapeutic use. This chapter is also concerned with the interests of women as these are affected by developments in new reproductive technologies. Specifically, this chapter deals with the question of whether doctors working in IVF practice are morally justified in asking women patients to donate some of their mature eggs, harvested in the context of treatment, for pre-clinical research. At this point, a brief introduction might prove useful for the non-specialised reader.

Human eggs are indispensable for pre-clinical studies into the feasibility and safety of egg manipulation techniques aimed at improving IVF treatment and allowing a broader range of patients to benefit from such research. For example, although cryopreservation of sperm and embryos is possible, cryopreservation of eggs is generally not: the problem is that there is no guarantee that the process will not cause damage to the egg and, as a result, make it useless or dangerous to employ in order to achieve a pregnancy. The difficulties concerning freezing arise because mature human eggs are very delicate as they are temperature sensitive and the disruption of the balance of their architecture may cause loss of chromosomes. Scientists, therefore, are working hard towards correcting this problem.

Moreover, as we saw in Chapter 4 in the context of developing embryonic stem cell therapy, human eggs are needed for studying the feasibility of therapeutic cloning through somatic cell nuclear transfer. Since it is mature eggs that are needed, their scarcity may inhibit the carrying out of that research.

As mature human eggs are obtained through donation, an ethical conflict arises between the duties of the doctor as physician and as researcher. We begin this chapter with exercises around a case study and paper by Wybo-Jan Dondorp.

The Case of Obtaining Eggs for Research: a Moral Bottleneck

Gynaecologists working in a University IVF center have plans for a preclinical study into the safety of freezing mature human eggs, in cooperation with researchers of their University's Department of Developmental Biology. Egg cryopreservation, if feasible and safe, would be valuable as a means to reducing the waste of surplus embryos in current IVF practice. As stated in the Netherlands Health Council Report on IVF related research, at present, in case there are many more eggs than the number of embryos that can be responsibly transferred, the options are either to produce and freeze supernumerary embryos or to allow the eggs to perish; usually the former option is chosen. The disadvantage of this is that supernumerary embryos can end up as surplus embryos. The possibility of freezing eggs could help avoid this. It would then be necessary to fertilise in each treatment cycle only the number of eggs needed to obtain the maximum number of embryos (two or three) that can be transferred at one time in the womb. Of course, because of the possibility of failure, more embryos would on balance be produced than the number transferred but still, the difference in number would be smaller than it is at present. This is a moral gain for those who consider the production of surplus embryos for IVF treatment unacceptable. Finally, it would yield a further (perhaps better) strategy of 'fertility insurance' for women undergoing aggressive oncological treatment (similar to cryopreservation of semen in order to protect a man's fertility from being wiped out by such treatment).

The envisaged study would encompass the following stages: obtaining mature donor eggs, cryopreservation, thawing, fertilisation and culture of the resulting embryos up to the blastocyst stage. The set up of the study would have to be such that it could be expected to yield better insight into fundamental biological mechanisms underlying the chromosomal damage that has frustrated earlier attempts (including clinical experiments) at cryopreserving human eggs in the sensitive stage of their meiotic development.

Some of the team are convinced that asking healthy women to undergo the necessary procedure is equivalent to recruiting healthy trial subjects in other branches of medicine. Others, however, believe that making such a request would be totally unethical. Nevertheless, all agree that this way of putting the issue does not bring them any further, since it is highly unlikely anyway that more than a few women would want to volunteer (let us assume in this case that donation for money is out of the question, as it is in the Netherlands and in other countries). (Dondorp 2000)

Activity:

The issue of healthy women, i.e., healthy volunteers asked to offer some of their eggs, raises specific concerns. Try to think what these concerns are and compare this situation with other kinds of research where healthy subjects are being recruited.

Now read an extract from Dondorp's paper on the same issue.

Eggs for research from IVF patients: the role of the doctor

Wybo-Jan Dondorp

Asking healthy women to subject themselves to the burdens and risks of hormone treatment and follicle puncture in order to donate mature eggs for research is not morally or legally unacceptable as such. The position of these women would be the same as (or at least comparable to) that of healthy subjects in medical research, which may also expose them to certain burdens and risks. However, if constraints on recruitment aimed at avoiding undue pressure on potential participants are heeded, only very few women can at best be expected to volunteer. In this context two important requirements are proposed:

1. That any reimbursements offered should not be so large as to invite prospective subjects to volunteer against their better judgement (Council For International Organizations of Medical Sciences [CIOMS]/World Health Organisation [WHO] 1993).
2. Moreover, it is accepted as procedural wisdom in recruiting volunteers for research that this should be done by a general notice rather than by directly approaching potential participants (Smith 1999).

An alternative approach would be to ask women of reproductive age undergoing gynaecological operations such as sterilisations permission for harvesting any mature eggs. Since these women have not undergone hormone stimulation, the maximum yield would consist of single eggs resulting from maturation in a natural cycle, thus limiting the usefulness of the procurement strategy. Thirdly, one could ask women having undergone IVF to donate any of their eggs that have failed to fertilise in the process. The cause of this failure not being known, a second attempt at fertilisation is usually not undertaken. The eggs being destined to perish anyway, there seems to be no moral objection in asking if they may be used in research. However, the reason for discarding these eggs in treatment may also make them less suited for the purposes of research. A final option is to ask IVF patients to donate some of the eggs harvested in the context of their own treatment and not yet exposed to a fertilisation attempt. Since there are often more mature eggs than the two or three embryos eventually to be transferred to the womb, it would seem that asking for any excess eggs would be perfectly acceptable. Or would it?

Of the four sources of mature eggs that Dondorp discusses:

1. healthy volunteers
2. women undergoing operations
3. eggs that have failed to fertilise in the IVF process
4. eggs that have not yet been fertilised in the IVF process

the most promising seem to be 3 and 4. Consider the following extract from Donna Dickenson's commentary on the case which addresses 3: the acceptability of asking women who have already undergone IVF to donate any eggs that have failed to fertilise.

Commentary on the case of obtaining eggs for research
Donna Dickenson

Since the woman is not being asked to undergo further follicle stimulation and egg extraction, but rather simply to 'recycle' the products of the procedures she has already undergone, this approach seems unimpeachable at first sight. The eggs will perish otherwise (unless, of course, donated to another woman which might be a moral imperative in Catholic countries). Much good could be done and there is no obvious harm. As we saw in Chapter 4, it is this kind of utilitarian argument that impelled the Nuffield Council working party on stem cell lines to conclude in favour of the use of surplus pre-embryos and embryos for stem cell research (Nuffield Council 2000). Since unfertilised eggs presumably carry less moral weight than fertilised blastocysts or embryos, to those who believe that embryos carry moral weight, the argument in favour of using discarded eggs this way seems even stronger. (Dickenson 2000)

Activity:

(a) From your experience and your reading so far, do you think that researchers and clinicians might be tempted to encourage a woman with a very low chance of successful fertilisation to undergo superovulation and extraction regardless, with a covert view towards using the eggs which fail to fertilise for their research?

(b) List any similarities or differences between sperm and egg donation. You might wish to review Chapter 2 on sperm donation. Do you think that a woman would be as willing to donate her surplus eggs to 'science' as she would be to donate them to another woman?

Case (continued)

The team agrees that the third strategy would yield a selection of possibly deficient embryos, something to be avoided in the envisaged study. This leaves only one alternative: asking IVF patients to donate a few of the eggs harvested after follicle puncture. According to some researchers in the team it is perfectly all right to make an appeal to the responsibility of those benefiting from IVF to contribute to research aimed at improving the treatment, so that future patients may also benefit. Others, notably the gynaecologists of the team are not so sure about that. They stress the fact that undergoing IVF is not the same as benefiting from it, and that, for these women, donation of their eggs may well reduce their own chance of having a child as a result of the treatment.

One of these team members quotes guidelines saying that it is only acceptable to ask IVF patients to serve as egg donors 'in exceptional cases, when a large number of eggs is obtained and a high fertilisation percentage is achieved in a prior attempt' and that 'no pressure may be placed on the woman to cooperate'. This, in turn, invites the comment that such conditions put a premium on the use of heavier stimulation protocols so as to obtain a larger number of eggs with a view to using some of them for research. Others ask whether the whole project is not better abandoned, given the impossibility of obtaining the required number of eggs in a morally acceptable way.

(The guidelines quoted here are taken from the report, *IVF Related Research*, 1998, of the Health Council of the Netherlands. The report adds to this that 'it is not acceptable to expose a woman to a heavier stimulation regime than is considered necessary for her treatment'.)

Activity:

Imagine that you had an opportunity to interview both the researchers and the gynaecologists at this point. Think of the following questions to the gynaecologists:

(a) Would you ask a patient of yours to donate some of her eggs for research? Justify your answer.

(b) If the answer is yes, would you feel that you are acting in line with your duty of beneficence towards your patient?

(c) If your patient refused to donate some of her eggs would this change your attitude towards her?

(d) Do you believe that your patients have a duty to benefit other couples through research?

Write down a list of possible counter arguments. Spend some time reflecting on this and then continue reading Dondorp's paper.

Contributing as donors, being affected as patients

No eggs can be qualified as 'excess' prior to fertilisation. Since the outcome of the process is hard to predict, fertilisation is normally attempted with all eggs obtained after follicle puncture. If this results in more than two or three good quality embryos, the excess number can be cryopreserved for later use by the same couple. Asking IVF patients to donate some of their eggs for research may, therefore, be tantamount to asking a sacrifice from them, consisting of a reduction of their own chances of having a healthy baby through the very procedure for which the eggs were obtained. Moreover, by reducing their chances of success they would also adversely change the balance of benefits and health risks connected to the undergoing of IVF treatment.

Activity:

Compare Dondorp's position with Dickenson's who makes a similar point in the following way.

Given the invasiveness and riskiness of the procedures which are involved in IVF treatment, can it be right to reduce the woman's chances of successful fertilisation below the rather miserable 16 percent which is currently the norm? (HFEA 1998). Let us translate this another way: is it right to knowingly expose women to additional, avoidable courses of superovulation and egg extraction? For the statistical group of all women who undergo IVF, even if not necessarily in the case of any individual woman, that will be the overall effect of allowing use of eggs before fertilisation.

Part of the moral difficulty in this case lies in the disparity between those women who are exposed to the risks of egg extraction and those who will (perhaps) eventually benefit. It might be argued that this is always the case with healthy volunteers in nontherapeutic research trials. But these women are different: if they are already subjects of IVF treatment, they occupy the misty borderland between the well and the 'sick'. (I assume that it would be more difficult to recruit women for the trial from the general population, and that the researchers' convenience would probably dictate a population from their own IVF clinic). We have been cautioned to beware of too readily accepting a medicalised model of IVF but infertility is construed as something other than perfect health in our society (Spallone 1989). These women are already vulnerable because they and/or their partners are infertile. They may well be readily susceptible to pressure from IVF researchers and clinics to accept participation in a clinical trial which will not benefit them directly, because they feel they 'owe' the researchers something, even if their own infertility treatment has failed.

Activity:

Whereas those participating as patients in a clinical study may often hope to benefit from doing so, IVF patients donating eggs for research have nothing to expect for themselves, as the consequences for the outcome of their treatment are neutral at best.

The fact that donation may affect the success or otherwise of treatment creates a strong presumption against doctors asking their patients. Do you think that those profiting from IVF (and thereby from earlier research) are under a moral obligation to contribute to research aimed at improving the treatment, so that future patients may profit as well?

Three arguments can be made which support the idea of doctors asking their patients to become donors: (1) that the patient is obliged to help others; (2) that it would make things different if the request were made by another person; (3) that the doctor could share her role conflict with the patient and let her decide. Dondorp considers each of these arguments in turn and finds each of them inadequate.

1. An obligation to help others

If we accept that an obligation to help others exists, wouldn't it be all right for the doctor to remind her patients of this obligation? After all, as a doctor she's committed to helping not just this patient but also others that may come for her help.

This will not do. Firstly, while helping others is always morally commendable, it can only be construed as a moral obligation in exceptional cases, when one's action would be needed to avoid significant loss or damage to a concrete other person, and when the probable benefit gained by that person outweighs any harms to oneself. In this case (donating eggs for research) neither of these conditions has been met. Nor would the fact that IVF patients profit from earlier research alter the situation. If there is such a thing as a duty of gratitude, this can only be understood as an 'imperfect duty' which would not lead to one's being morally obliged to perform a particular act of beneficence (O'Neill 1991)

Secondly, it is still uncertain whether the prospective donors will themselves profit from undergoing IVF treatment. Asking them to consider donating some of the eggs to pay for what they may or may not gain is to run ahead of things. Moreover this reasoning is quite inconsistent because donation, if it would occur, would actually decrease the chances of probability of a successful IVF treatment of the candidate parents.

Thirdly, even if IVF patients would accept for themselves some sort of moral responsibility to contribute to further infertility research, this would not suffice to justify any requests to that end put before them by their doctor. In connection with the fact that donation or refusal of donation may affect treatment, the possible presence of such a motivation, should, in contrast, be seen as a reason for even greater reservation on the part of the doctor.

Activity:

Dondorp makes three objections to the claim that doctors are justified in asking their patients for egg donations on the grounds that patients are obliged to help others. Briefly list his three objections. Do you agree with each of them? Can you think of any others?

2. Role conflict and role differentiation

Doctors profess that the health of their patients is their first consideration (WHO, Declaration of Geneva 1948). As happens with all professional bodies, the role of such a pledge is to engender public expectations concerning them (Koehn 1994). These expectations are the basis on which patients entrust themselves to the care of those belonging to the medical profession. For maintaining the moral fabric of the doctor-patient relationship it is essential that this trust be preserved (Sokolowski 1991).

However, doctors also act in other capacities or roles, which may occasionally conflict with their primary role as doctors. One such capacity is that of researcher. While, as researchers, doctors also serve the good of health that is the moral end of their profession, it is not the health of their individual patients with which they are primarily concerned in that capacity. The situation we are considering is a good example of the kind of conflict between role requirements that may occasionally arise. It is in the interest of the kind of medical research in which IVF doctors may be involved that more mature human eggs are donated for the purpose of such research. Given the limited yield to be expected from alternative procurement strategies, this might lead them to consider asking their patients. On the other hand, the fact that donating any mature eggs may reduce their patient's chance of having a baby through IVF, would make it very difficult for them to justify such a request from the perspective of what it means to be a doctor.

One proposal to handle such tensions involves role differentiation.

Role differentiation: the role of the person responsible for research activities is separated from the role of the treating physician.

A recent example is the requirement in the US National Institutes of Health (NIH) *Guidelines for Research Using Human Pluripotent Stem Cells* that, in order to achieve 'a clear separation between the decision to create embryos for fertility treatment and the decision to donate human embryos in excess of clinical need for research purposes...the attending physician responsible for the fertility treatment and the researcher or investigator deriving and/or proposing to utilise human pluripotent stem cells should not have been one and the same person' (NIH 2000). Similarly, in the context of research using tissue from aborted foetuses, separating the roles of those responsible for the transplantation and for the abortion has been suggested as an additional measure to ensure that the woman's decision to terminate her pregnancy will not be influenced by the tissue's potential use for transplantation (Boer 1994).

However, an important difference between the situation we are considering at present and the two examples just cited is that, even if the roles were separated and the research was being done by a person other than the treating physician, the problem would not be solved: the donation of eggs, no matter who the responsible researcher is, *does* affect the outcome of current treatment for which the doctor remains fully responsible. Letting another member of the research team do the request would therefore not solve the conflict. Patients would also want to discuss the implications of the request with their doctor. What would be her advice to them? Short of taking the doctor off the research in question, role differentiation would therefore not make much of a difference.

Alternatively, to take the doctor off the research and allow her to partake in similar studies, though not with her own patients, would at least restore her to full advocacy of her own patients' interests. However, role conflict would remain if, as a doctor, she partakes in research for which her colleague's patients are recruited as egg donors on terms that she would not find acceptable were she herself responsible for their treatment.

Further, from the doctor's non-involvement in the relevant research project, it does not follow that she could turn a blind eye to any of her colleagues addressing her patients with a request that might affect the outcome of their treatment. Being responsible for the treatment means that she would have to protest against any such actions planned or undertaken by a colleague. Failing to do so would make her responsible for the request.

We may conclude that because of the treatment-affecting implication of egg donation by IVF patients, role differentiation would not, from a moral point of view, make it any easier to ask them.

> ## Activity:
>
> The issue of role conflict is not limited to professionals working in IVF clinics but is also an issue for GPs, psychiatrists and nurses who are often asked to recruit patients for clinical trials. In these cases, their research commitments and their duty to promote what is in the best interest of their patients involve them in conflicts which they must resolve. Does your experience in any of these areas throw any further light on the issue of role conflict in the case of egg donation?

Role differentiation has been proposed as a solution to the problem of role conflict but apparently it does not work very well. This is because even if the professional roles of researcher and treating physician are separated, the donation which has taken place still affects the outcome of the IVF treatment. Alternatively, think about 'sharing role conflict with the patient'. This is a different approach outlined by Dondorp below.

3. Sharing role conflict with the patient

This proposal for handling role conflicts in professional-client relationships involves providing full disclosure of possibly competing commitments (Bayles 1988). The question is not whether disclosure of the doctor's involvement in the research should be an element in the information given to the patient whenever a request for a donation is made. It is clear that it should (Beauchamp and Childress 1994). The question is rather whether providing the patient with this information would justify a request that is clearly not in the patient's interest. If it has been made clear that the request is made not by the doctor *as doctor*, but by the doctor *as researcher*, the fact that it would not be in the patient's interest to donate any of her eggs for research can be left to the patient to consider. Or so the argument would go. The idea behind it would be to solve the conflict by sharing it with the patient and let her decide. After all she may find it important to be given the possibility to contribute to research from which other patients may benefit, or at least to weigh the possibility against a consideration of her own interests in the case.

The problem with this suggestion lies in the presupposed view of the doctor-patient relationship and its moral basis. It is seen as nothing but a contractual relationship between two fully autonomous partners agreeing on the deliverance of services under conditions accepted by both sides of the contract. On this view, it is the fact of the contract not the nature of the services (medical help) that determines the morality of the relationship (Veatch 1981, 1988). This view is problematic because:

- it misrepresents the meaning of these services for those (all of us) who depend on them for their health and well-being (Pellegrino 1983) and
- it ignores the importance of professional power and patient vulnerability as essential features of the clinical encounter (Lebacqz 1985)

In the case we are considering, IVF patients are in a very vulnerable position. For many infertile couples, IVF treatment offers a last chance after a history of hopes and deceptions, during which their involuntary childlessness has often grown into something close to an all-absorbing obsession. In this situation they have entrusted themselves completely to the doctor and her team. If this doctor then tells them she is also a researcher and suggests that they opt for that research and perhaps donate some of their eggs which may benefit many other infertile patients, a positive answer can be expected. However, chances are real that this would be the result, not so much of an autonomous weighing of pros and cons, but of the all too understandable fear of falling out of grace with the one person who may change their fate. What this suggests is that the request may have a trust-undermining effect, which can as such not be accounted for in the contract model of the doctor-patient relationship. If this analysis is correct, the idea of solving role conflict by sharing it with the patient cannot be maintained.

Conclusion

The above discussion leads to the conclusion that the option of asking IVF patients to serve as egg donors could only be considered in exceptional cases, when there is a reasonable degree of certainty that donating a small number of eggs would not lead to a meaningful reduction of the patient's chances of success. This would be the case when a large number of eggs is obtained and a high fertilisation percentage is achieved in a prior attempt (Health Council of the Netherlands 1998). These being exceptional cases, the option of asking IVF patients would indeed only marginally help to diminish the shortage of eggs for research.

In the exceptional cases meeting the description just specified, a donation request could be acceptable under further conditions. It goes without saying that there would be no justification for exposing the patient to a heavier stimulation protocol than would be needed for her treatment, so as to obtain a larger number of eggs and then be in a position to make the request. Furthermore, if the eggs are meant to be used in research for which the approval of a medical ethics committee is required, this must be obtained prior to asking the patient. Moreover, the patient (and her partner) ought to be given all information required for making an informed decision (American Society for Reproductive Medicine

[ASRM] 1997). This would include disclosure of the doctor's involvement in the research and the reasoning underlying the judgement that donation would not lead to a meaningful reduction of the couple's own chances of success. Also, information should be given about the nature of the research and whether or not it might include the creation of embryos. Prospective donors should be given the option of selective consent for research not involving this implication and be informed of the possibility of retracting their consent at any time until the actual start of the study. Finally, prospective donors must be assured that nonparticipation will not adversely affect their status in the treatment programme.

Activity:

(a) Dondorp argues that 'there is no justification for exposing the patient to a heavier stimulation protocol than would be needed for her treatment'. Can you think of any reasons why he makes this claim? What principles, central to the doctor-patient relationship, are at risk of being infringed here?

(b) Dondorp concludes that the option of asking IVF patients to donate some of their eggs can only be considered in exceptional circumstances, and only then, under strict conditions. Do you agree or disagree with his conclusion? If you agree, are there any further conditions that you think should be put in place which would protect the health, well-being and interests of any patient donating eggs? If you, yourself, were a woman patient attending an IVF clinic would you consider donating eggs that had

- failed to fertilise in the IVF process?
- not yet been fertilised in the IVF process?

Summary

This chapter explored the ethical issues connected with the conflicting role of the IVF clinician, as doctor and researcher, in relation to the donation of human eggs. We discussed the relevant obligations and responsibilities in relation to donation and the doctor-patient relationship. Further, we assessed

the impact of donation on the health, well-being and interests of women, on the success or otherwise of IVF treatments and on the continuing development of NRTs. In particular, we
- examined the distinction between different sources of mature human eggs
- discussed the implications of conflicting professional roles
- assessed the risks and benefits of egg donation for all parties involved

Who Decides?
(Access to IVF Treatment)

Objectives

At the end of this chapter you should be able to:

- distinguish between the right to procreate and the right to medical assistance in procreation
- consider the possible conflict between respecting the autonomy of candidate parents and ensuring the welfare of the child-to-be
- explain the maximum welfare principle, the minimum welfare principle and the reasonable welfare principle and apply them to specific case studies

In each of the previous chapters of this workbook, the issue of access to new reproductive technologies (NRTs) has been touched on in various ways. This is because the question of who should and who should not have access to the techniques of assisted reproduction lies at the heart of the debate concerning the ethical dilemmas raised in this context. It is closely related to different moral norms and values (such as respect for patient autonomy, responsibility of the doctor or nurse) as well as to various and often conflicting interests. It is, moreover, related, to the issue of what the goals of medicine are.

In 1988, a Report of the Council of Europe dealing with the issue of assisted reproduction stipulated that NRTs should be used only when other methods for the treatment of infertility had failed or were not appropriate in the particular case; they should not be used as alternative means of procreation. This Report reflected the conceptualisation of infertility as a permanent disability, functional impairment or handicap which requires medical treatment (Holm 1996). Infertility, however, is not a condition that causes physical but rather psychological and social suffering. Thus, the treatment offered is not therapeutic, as the majority of methods of medically assisted reproduction do not eliminate the causes of infertility and consequently do not treat it; they rather 'cure' childlessness. However, the introduction *per se* of IVF techniques has led to the conclusion that although medical treatment in the strict sense does not take place, it is accepted to use other means (medical procedures) to assist the conception of a child. Infertility is seen as a health need (although this assumption is not accepted by all those who experience it) which can be satisfied through various medical procedures (Evans 1990, Martin 1996).

The goals of medicine

According to Dorland's Medical Dictionary, medicine can be defined as

> the art and science of the diagnosis and treatment of disease and the maintenance of health. (Dorland 1994: 999)

This conventional approach to medicine has been challenged by the results of a project regarding the 'Goals of Medicine' carried out by the Hastings Center (1996). In this project it is stated that medicine is constantly changing, not only at the clinical level, but also at the level of biomedical research and economics. However, it is argued that these two views of the nature of medicine and its goals complement each other even as they contend with each other; one discerning inherent goals, the other discovering only time and culture bound socially constructed goals. It is true that there is great variation over time and in different cultures in the nature and goals of medicine. The main point of the thesis of this project is that the interpretation of disease is so varied and the response to illness and sickness so complex that it is difficult to pin down a meaningful set of inherent values and convictions. The importance of the inner direction and core values of medicine is emphasised but it is stated that it would be naïve to think that medical values can remain uninfluenced by society.

One of the fields which best reflects these diverse approaches to the goals of medicine, is the field of NRTs. According to the Hastings Center Project, one of the goals of medicine has been the promotion and maintenance of health and infertility has indeed been defined as a health need. However, considering the complex nature of these goals, can we assume that childlessness, even when it is not caused by infertility, could, in our society take the form of a psychological need that should be alleviated through medical means?

Activity:

Stop here for a moment and consider what you think of as the goals of medicine. Make a list and spend some time reflecting on it. Compare it with the paragraphs that you have just read and try to summarise those features which you think are important elements of good medicine in the field of NRTs.

Now read the following Greek case study presented by Panagiota Dalla-Vorgia and Tina Garanis-Papadatos. This case reflects new possibilities for the role of medicine: Diana is a woman who is not infertile but childless, a woman who chooses to undergo IVF for her own personal and not medical reasons.

The Case of Diana

Diana, a forty-two-year-old single woman, appeared at a private IVF clinic with the aim of having a child by donor insemination. Diana has studied both in her home country and abroad and has excelled in her professional life. Nevertheless, she never had time for a family life as she always believed that serious relationships entailed a kind of commitment that she was not willing to undertake. Lately, however, after a successful career and some brief relationships with men, she has started sensing a feeling of loneliness and failure. She realises that a child is all she wants. It is at this very moment that she seeks professional help in order to conceive.

The first two attempts to help Diana conceive a child failed, but the third one was successful and in December 1997 Diana was holding her baby boy in her arms for the first time. Her family had been very supportive from the beginning. 'I know', Diana says, 'that no matter how much I protect my son, there will be a moment when he will start asking questions and when some people are most likely going to hurt him. But he will be prepared for this. My son will never have a father as the sperm donor is anonymous. Nevertheless he has been given the gift of life. He also has a family: grandparents who adore him, a loving godfather, uncles and cousins'.

Diana is working closely with a child psychologist, preparing her steps one by one. She is convinced that one day her child will be in a position to understand the acts of his mother and to forgive – if there is anything to forgive at all. (Dalla-Vorgia and Garanis-Papadatos 2000)

With the advancement of reproductive technology there has been a growing number of women requesting fertility treatments in order to have a child without a male partner. One of the reasons for this is that they see the use of donor insemination as preferable to sharing parental responsibility with a man with whom they are not emotionally involved (MacCallum 2000b). In Greece, as there exists no specific legal framework, it depends on the doctor's discretion whether he will accept the woman as a candidate. In the UK it is legal for a single woman to receive fertility treatment. However, fertility clinics are not obliged to treat these cases and only about 10 percent choose to do so. This is due to s.13 (5) of the HFE Act (1990), which requires the clinic to take into account the welfare of the potential child, including his [sic] need for a father. Thus, a single woman in Edinburgh requesting donor insemination may find herself having to fly to London for a clinic that will accept her. This situation may change with the recent incorporation of the European Convention on Human Rights into UK law, which could allow women who are refused treatment by their local clinic to challenge the decision in the courts.

Activity:

What is your opinion of the following statement: 'Diana feels entitled to receive the professional help she needs in order to conceive a child; she feels she has the right to become a mother like any other woman'. This raises the issue of the 'right to procreate'. Do you believe that such a right exists? The easy answer would seem to be 'yes'. Try however to think what this would entail. Do you think that the decision to procreate is a purely personal one?

If we accept that a natural right to procreate does indeed exist, it is possible that we also accept the fact that people should never be prevented from procreating and, moreover, that they should be helped to procreate, should they wish to do so. According to Martyn Evans this would raise questions of access to health care services and entail obligations 'of the most dramatic kind' (Evans 1996). Evans goes on to say that although the basis of procreation is biological, the desire for children is a social phenomenon, a fact which could be interpreted in many different ways: the right not to be prevented from having a child, would not entail the right to be provided with a child. In this context, Dooley points out that

> few decisions are solely personal and the decision about parenting or motherhood are affected by causal influences – familial, religious and social – which may be subtle but are real and felt and anything but simple. A shared concern by many feminist writers is that techniques such as in-vitro fertilisation coexist with a powerful ideology of motherhood as a biological imperative rather than a social relationship. (Dooley 2000)

While it is true that Article 8 of the European Convention on Human Rights refers to the right to private and family life and Article 12 protects the right to marry and found a family, these provisions do not entail the existence of an absolute right of access to infertility treatment. Although most European countries have signed the European Convention, legislative approaches differ between them. The discrepancies in views can be explained by their differing commitments to underlying ethical concepts such as autonomy and human dignity. Thus more or less liberal approaches can be seen as results of 'divergent basic ideological questions: "what should be allowed?" in Austria and Germany, *vis-à-vis* "what should be prohibited?" in England' (Bernat and Vranes 1996).

Now that you have considered one case involving access to infertility treatment, we would like you to read a paper by Guido de Wert and Ron Berghmans (2000b) which focuses on the issue of access to treatment by particular individuals/couples where one or both partners is/are HIV-seropositive or where one or both partners pose(s) a genetic risk to their offspring, an issue to which we will come later. For the moment we will concentrate on their views regarding the right to procreate in general.

In Vitro Fertilisation: access, responsibility of the doctor and welfare of the child

Guido de Wert and Ron Berghmans

Should doctors honour any request for assisted procreation? Is the doctor's sole responsibility to assess medical issues, and if patients are medically suited for treatment to conduct IVF in conformity with the state of the art? Or have doctors a specific responsibility that goes beyond respect for the autonomy of the woman/couple that asks for assisted reproduction? Given the fact that patients need the assistance of doctors in reproduction, doctors cannot deny moral responsibility for their actions. As a matter of fact, the possible child that comes out of their intervention would not have existed without the involvement of the doctor in their coming to existence.

Some argue that it is not justified to deny access to IVF treatment on grounds other than medical suitability. They claim this to be an unacceptable violation of the principle of respect for patient autonomy and of the internationally proclaimed 'right to procreate'. They fear ideological preoccupations as well as ungrounded prejudices of the doctor. Also, it is often argued that people who can procreate 'normally' are not judged beforehand and do not need a 'licence to procreate'. Against this categorical claim we defend the view that doctors have an obligation to consider the interests of the future child. This means that respecting the autonomy of the parents-to-be is an important *prima facie* principle, but not an absolute one. The freedom of procreation, or the right to procreate cannot be claimed as a foundation of the right to medical assistance in procreation. Although a possible right to assisted procreation cannot be solely grounded on the right of procreation, this does not mean that a right to assisted procreation does not exist, but that its foundation lies elsewhere (possibly in a duty to benefit others). We defend the view that doctors have a role-specific responsibility in the context of assisted reproduction. This role-specific responsibility concerns the interests of the child-to-be.

Activity:

Stop here for a moment and propose some criteria for evaluating the future welfare of the child-to-be. The welfare of the child has been characterised as 'the most complex issue to assess scientifically and medically because of its largely psychosocial component' (IFFS 1998). To what extent do you think that the welfare of the child-to-be is compatible with the parents' absolute autonomy?

As Fiona MacCallum (2000b) points out, there is no clear consensus on how the child-to-be's welfare should be evaluated. Some would argue against allowing access to infertility services to anyone other than traditional married couples, with the belief that this is the 'ideal situation' and that all other family types are inferior in some way. Early research on single-parent families found that children did do less well than in two-parent families in terms of psychological adjustment and academic achievement. However, these were women who had not chosen to be single parents and there were contributory factors present such as economic hardship and the emotional consequences of the break up with the child's father. In comparison, the newly growing group of 'solo' mothers by choice tend to be financially secure and their children, raised in a single family from the beginning, are not exposed to the conflict and distress of separation or divorce. Evidence suggests that in these families, positive outcomes for children depend on the mother being economically stable and having a strong social support network. In the case study Diana seems to have both these elements with a successful career and a loving family (MacCallum, 2000b).

Returning to de Wert's and Berghmans' paper we will see how they attempt to evaluate the welfare of the child-to-be using three different criteria.

So, an important question concerns the future welfare of the child that is the 'product' of assisted reproduction. Everybody agrees on the fundamental importance of the welfare of the child when judging the applicability of new reproductive technologies. But what evaluation principle should guide the interpretation of the level or measure of welfare of the child that is the result of assisted reproduction?

Three principles, that can be seen as positions on a continuum between the absolute autonomy of the parents on the one hand, and exclusive concern for the welfare of the offspring on the other, can be distinguished: the maximum welfare principle, the minimum threshold principle and the reasonable welfare principle (Pennings 1999).

The *maximum welfare principle* implies that one should not knowingly and intentionally bring a child into the world in less than ideal circumstances.

Since we can control (at least to a certain extent) the circumstances in which a child is made when the candidates are infertile, we ought to restrict our cooperation to those cases which maximise the welfare of the child. At the same time this fact explains why the standard for medically assisted procreation must and can be higher than for natural reproduction (Pennings 1999).

On the basis of this principle, the existing child is harmed to the extent that he or she could have had a better life. This is an extremely strong standard, and when we consider the consistent application of this rule we will have to conclude that the overwhelming majority of the population ought to be excluded from procreation.

The *minimum threshold principle* is connected to a weak interpretation of the right to procreate: 'a person has a right to rear children if they meet certain minimal standards of child rearing'.

This standard does not compare the welfare of the child-to-be-created with other possible children but it only verifies whether the quality of life of the future child is above the minimum threshold.

The concrete circumscription of all dimensions of the welfare of the child proves almost impossible. This problem is closely linked to the fact that we have not yet been able to formulate adequate criteria of a good parent. However, experience in law has taught us that a reasonable consensus can be reached on which conditions or characteristics of the parents are unacceptable. A survey of fertility centres in the US revealed that four criteria were used for treatment rejection: substance abuse, physical abuse, severe marital strife and coercion of one spouse by another.

Still, even this minimal threshold may be the subject of disagreement. The field of medical genetics presents several examples of fundamental discord on the right to procreate of parents who have an increased risk of having a child with serious malformations. A similar debate is going on concerning the rights of parents infected with human immunodeficiency virus (HIV).

One of the most frequently used minimum threshold standards can be called the 'wrongful life' or the 'worse than death' standard: A child should not be brought into the world if it would be better for him or her not to have been born at all (Robertson 1994). This is an extremely low threshold: if every act of procreation that meets this standard is considered acceptable, we cannot even say that it is wrong to bring a child into existence whose prospects and opportunities are awful.

The *reasonable welfare principle* refers not to the perfectly happy child but to the reasonably happy child.

The reasonable welfare principle is an intermediary principle between the maximum welfare principle and the minimum threshold principle which conforms more closely with the way we look at procreation and parental responsibility in ordinary life (Pennings 1999).

On the one hand we do not have to reject or criticise people for bringing a normal child into the world because they could have had a happier one. On the other hand, we are not forced to accept decisions which result in the birth of seriously handicapped children because the net result is a life still worth living. The amount of welfare of the child may be lower than the level that could be expected in ideal circumstances but it may still be optimal given the concrete circumstances and characteristics of the parents.

Activity:

What do you think of the above mentioned principles? Would you have a preference for any of them? On what grounds is that preference based?

HIV and IVF

Given the relevance of the responsibility of the doctor for the well being of the future child, de Wert and Berghmans refer to the example of access to IVF treatment for HIV-seropositive women/couples in order to tease out some of the issues that they have been discussing. They point out that testing for HIV in the context of assisted reproduction is not a new phenomenon. It is, for instance, a requirement of professional practice to test (candidate-)donors of sperm or eggs, for HIV. Those who refuse testing or are seropositive are not accepted as donors. According to the authors, in general, a doctor who is confronted with a couple that wants assistance in reproduction will not be informed about the HIV-serostatus of the couple. The physician will bring up the issue of HIV, firstly, in the context of informing the couple about risk factors for HIV and, secondly, about the risks of HIV for the offspring.

Activity:

Imagine the following situation: you are the doctor or clinician interviewing the couple of candidate parents.

● Would you ask them directly about their HIV status?
● Would you have a feeling that you are violating their autonomy by asking these questions?
● Would a negative answer from the people in front of you regarding the fact of being seropositive or not, affect the way you would act?
● In case one or both candidate parents prove to be seropositive and access is denied to them, would you feel that you are overriding your duty to benefit your patients? How would you justify your answer?

Now, read the views of de Wert and Berghmans on this issue and consider your response to them.

The question of whether or not doctors, taking into consideration the responsibility for the welfare of the child-to-be, could or should deny HIV-seropositive candidate parents access to IVF treatment, cannot be answered in a clear-cut sense. Firstly it makes a difference *who* is infected: only the woman, only the man or both? If only the man is infected, then sperm isolation and washing procedures can significantly reduce HIV levels in the inseminating fraction (Anderson 1999). If the woman, or both candidate parents are infected, then the risk of vertical transmission from mother to infant is still about 10 percent. In this last case, there is also the strongly increased risk that the child will become an orphan at an early age. A second reason to differentiate concerns developments with regard to the prevention of vertical transmission and of better medicines for HIV-seropositive patients. This implies that the medical as well as the psychosocial risks for the future child are becoming smaller. This also means that giving HIV-infected candidate-patients access to IVF treatment cannot be considered *a priori* morally unacceptable.

Requesting HIV testing as a condition for IVF treatment is at odds, for example, with the principle of respect for patient/client autonomy. This policy nevertheless can be justified on the ground of the role specific responsibility of the reproductive physician to take into consideration the interests and the welfare of the child-to-be. Although the positive developments with regards to reduction of the risk of vertical transmission and combination therapy may justify in individual cases a more permissive access policy, they nevertheless do not justify abandonment of a policy of offering the HIV test as a condition for access to IVF treatment. Offering access to IVF treatment to infected candidate-parents, because of the specific medical treatment and other arrangements that are needed, implies that the doctor is informed about the serostatus of candidates who are at high risk, and whether they are able and willing to comply with necessary arrangements.

Activity:

In the light of your own experience and the practice in your country, begin a list of the various advantages and disadvantages of testing the serostatus of candidate-parents. Then proceed to reading Fiona MacCallum's commentary on the de Wert and Berghmans position.

Commentary on '*In vitro* fertilisation: access, responsibility of the doctor and welfare of the child'
Fiona MacCallum

In the UK, there is no nationwide policy on HIV screening as part of the assessment of patient's suitability for IVF treatment. Individual clinics are free to devise their own protocols for testing and for management should the test prove positive. A recent study of IVF clinics showed that policies ranged from seeing HIV screening as essential (38 percent of IVF centers) to regarding it as unnecessary (24 percent), (Marcus et al. 2000). One clinic where all patients are screened argues that every clinic should do the same, basing this contention on the fundamental importance of the future welfare of the child. If one adheres to the maximum welfare principle, there is a case for universal screening of all prospective IVF patients. Admittedly for the large majority of these patients the risk of infection is small, but strict application of the maximum welfare principle means that any risk at all should be ruled out if possible. Conversely on the basis of the minimum threshold principle, particularly if the threshold is set at the level of the 'worse than death' standard, it could be argued that there is no need for HIV screening of any patients. The possible consequences of an undiagnosed HIV- positive parent for the welfare of the child are severe, in terms of both the child's own health and the possibility of losing that parent at an early age. However, even in this situation it is not necessarily justifiable to state that it would have been better had the child never been born.

The less extreme requirement of the reasonable welfare principle is that one ensures the optimal welfare of the child 'given the concrete circumstances and characteristics of the parents'. Using this standard might lead to selective screening, where only those for whom the risk of infection is substantive are tested. Grounds for selection will be difficult, but not impossible to formulate. Application of these criteria may be the greater problem since, to a large extent, clinicians will have to rely on the patient's self report of their participation in high-risk behaviour. Patients applying for IVF treatment may not always be forthcoming with full details of their lifestyle since they are worried that it may adversely affect their application. As a result, some HIV-seropositive patients would not be tested. There is also the possibility of offending or distressing those patients who are selected for screening and test negative for HIV. As Marcus et al. (2000: 1657) point out 'routine screening causes less anxiety to the couple than selective screening'. Universal testing would be the best policy if clinicians only had to consider the future welfare of the child but they also have a duty of care to the couple.

The situation becomes more complex when one takes into account the moral requirements of the patient–doctor relationship, such as the duty of beneficence/ nonmaleficence of the doctor. Although HIV testing is not incompatible with the duty of beneficence, in principle, routine screening might be seen to conflict with the principle if candidates who tested positive to HIV tests were automatically denied access to treatment. The UK Department of Health has recently advised that all pregnant women be offered an HIV test to minimise risk of transmission to the foetus. In line with this, the nondiscriminatory option would be for IVF clinics to treat their patients in the same way and screen for HIV only after the establishment of a pregnancy. However, this would create a conflict with the doctor's professional and legal responsibility for the child's welfare. A better option would be for each couple's case to be considered individually rather than imposing a blanket ban. The risk to the child of having a seropositive parent is not uniform but depends on factors such as which parent is infected and what stage the viral process has reached.

The need for a coherent policy on the screening issue across IVF clinics in the UK is clear since the current situation lays open the possibility that a couple who are aware that they may be at high risk will choose to apply to a clinic which does not include HIV screening in its protocol. Whether one is in favour of IVF treatment for these couples or not, knowledge of the HIV status would allow appropriate counselling to be given and informed decisions to be made by all parties.

Activity:

Below we have compiled a table of some of the arguments for and against testing for HIV that have been presented in this Chapter. Take a few minutes to consider the list of arguments and see if you can add anything further to it.

Arguments for and against testing the serostatus of candidate parents

For	Against
✔ HIV testing is consistent with the responsibility of the reproductive practitioner to consider the interests and welfare of the child-to-be	✘ requesting HIV testing as a condition for IVF treatment disrespects the autonomy of parent-candidates
✔ in cases where tests prove positive, preventive medicine can reduce vertical transmission of HIV	✘ the test outcome affects the way candidates are treated by practitioners on different levels and may mean that they are refused treatment

✔ the possible consequences of an undiagnosed HIV-positive parent for a child are severe: his or her own health will be threatened and they may lose a parent at an early age

✔ if candidates who tested positive were not automatically denied treatment, then the testing could be perceived as a benefit to all concerned

✔ ?

✗ on the minimum threshold principle, the life of a child with severe disadvantages, such as those envisaged here, is still preferable to no life at all

✗ people who need medical assistance with reproduction should not be treated any differently to others: if people generally are not required to take HIV tests, then these people should not be obliged to take the tests

✗ ?

Activity:

We will now come back to the case of Diana, or rather, a variation of it involving a forty-two-year-old lesbian, Linda. Please read the case and jot down your immediate response to the ethical issues that arise. How do you think the case compares with Diana's?

The Case of Linda

Linda is forty-two years old and involved in a stable relationship with another woman for the past three years. Like Diana, she is very devoted to her work. Realising that assisted reproduction is the only possible way for her and her partner to have a child, Linda visits a private clinic, asking for help. She does not hide the fact that she is a lesbian, on the contrary, she and her partner visit the doctor together. The latter refuses to accept Linda as a candidate, claiming that helping a lesbian to create a family is against his personal beliefs regarding the traditional family nucleus.

Discussion

Like Diana, Linda is not infertile but childless. In both cases, the traditional family pattern is lacking. However, as MacCallum (2000b) points out in her commentary on this case, with a lesbian mother there are concerns beyond the absence of a father. Some assumptions may be made, e.g.,

a) the child will be stigmatised by peers and others in society due to the stigma still attached to homosexuality

b) gender development will be adversely affected by the unconventional role models surrounding the child, and

c) the child will grow up to be homosexual which is seen by some as undesirable

Activity:

Stop here for a moment and try to find possible counterarguments to these assumptions.

McCallum points out that all the research on children raised by lesbian parents has shown that there is no support for any of these arguments and that the families are functioning normally: these children fare as well as those from traditional families in terms of social, emotional or gender development (Brewaeys et al. 1997, Golombok et al. 1997, Chan et al. 1998). In Linda's case the participation of her partner can be seen as positive since in two-parent lesbian families, the co-mothers seem to be more involved with the children than are the fathers in heterosexual two-parent families. In general, family processes such as quality and style of parenting have been shown to be more important for a child's psychological well-being than family structure such as number of parents or mother's sexual orientation.

According to de Wert's and Berghmans' reasonable welfare principle, if we accept lesbians or single older women for infertility treatment, we do not have to deny that heterosexual parents are better or that younger women are more competent parents. The reasonable welfare principle is not threatened if this ever could be shown to be true. We only have to show (and the evidence at the present time is already very convincing on this point) that they make acceptable and suitable parents. The amount of welfare of the child may be lower than the level that could be expected in ideal circumstances but it may still be optimal given the concrete circumstances and characteristics of the parents.

> The use of the reasonable welfare principle implies that conception at a certain point in time and in certain circumstances might be worse than conception at a different time and in different circumstances, but that does not entail that the decision is unacceptable. In other words: the welfare of the child might not be as high as it could have been (i.e., had the child had different parents) but it is sufficiently high to be considered a positive gift to the child.) (deWert and Berghmans 2000b)

Activity:

(a) Reflect on the suggestion that the 'ideal circumstances' for parenting can be those in which the parents of children are married, heterosexual and young. What is your own response to this?

(b) What do you think would be the outcome of this case in your country? Write down a set of conditions which you think should apply to any cases involving access to IVF.

(c) Do you think that the doctor in this case should be permitted to object to treating Linda on conscientious grounds?

The moral evaluation of the lifestyle of potential parents is a thorny issue as questions of fairness and impartiality are being raised. What role should the doctor's individual moral beliefs play?

> Certainly if whole classes of people were to be objected to, e.g., lesbian mothers, mentally handicapped and single women, much more evidence to establish the claim that members of such groups would make bad parents would be called for – and I suspect be impossible to find. In individual cases a history of child abuse, violence and instability may be much better grounds for such objection. (Evans 1990: 16)

One should think here of the implications of accepting the fact that doctors have a right to refuse to carry out an intervention which is contrary to their basic moral beliefs. In English medical law there exist two clauses which allow practitioners to refuse to treat a patient on other than clinical grounds, the first concerns abortion and the second concerns medically assisted reproduction. The latter concerns practitioners who object to assisted

conception in principle, or practitioners who object only to particular cases, i.e., cases where nonmedical criteria of suitability
would apply (Evans 1990). (See Appendix 1 for a breakdown of
the eligibility criteria for access to medically assisted reproduction
across the European Union.)

Summary

This Chapter looked at some of the ethical problems that are
related to access to medically assisted reproduction. Specifically, we

- discussed the notion of the right to procreate and the right
to medical assistance in procreation
- examined the potential conflict between the candidate-
parents' autonomy and the child-to-be's welfare
- introduced and applied the maximum welfare principle, the
minimum welfare principle and the reasonable welfare principle to particular cases

– 8 –

Social Responsibility
Liberty and the Public Good

Objectives

At the end of this chapter you should be able to:

- discuss the concept of social responsibility in the light of ethical challenges raised by reprogenetics
- weigh arguments for and against preimplantation genetic diagnosis (PGD) in the light of individual liberty and the public good
- decide whether or not genetic testing contributes to or promotes discrimination on disability or gender grounds

> Do not let our capacity to do, blind us to our responsibility only to do good
>
> (Silver 1999: 68)

The focus on social responsibility in this final chapter brings us full circle from Chapter 1 where we introduced the concepts of professional and parental responsibility. This third dimension of responsibility expresses the idea that in addition to special obligations ensuing from professional and parental roles, the notion of *responsibility is significantly linked to social relationships with others who are not in close relationships with us but are citizens of our society or our cultural community.* Alastair MacIntyre argues that the interpersonal and social contexts of any happening are pivotal in understanding responsibility.

> Conceptions of responsibility are never just thoughts in the head:
> they are actual or potential modes of practical engagement with
> others, embodied in social institutions. So, a choice between alterna-
> tive conceptions of responsibility is a choice between different types
> of social relationship and different types of institutions. (MacIntyre
> 2001)

If MacIntyre's insight is correct, then the choices to be made
about New Reproductive Technologies (NRTs) by members
of a society will be choices about the types and qualities of
social life and relationships they wish to develop. These
choices would then have consequences for the types of institutions (family, health care, education, etc.) which might best
effect this quality of social life. In this chapter, building on
MacIntyre's insight, we ask: In a given society, when provision is made for a range of choices in the use of NRTs, are
these choice provisions, at the same time, decisions about the
types and qualities of social life and relationships that we
wish to develop? We think the answer to this initial question
is 'yes' but the extracts, case and arguments of this chapter
are intended to illustrate the plausibility of that viewpoint.

By way of review, before moving into this task, look back
at Chapters 2 through 7. There you examined case studies
reflecting a range of choice options under the umbrella of
reprogenetics: IVF and PGD decisions, donor insemination,
surrogacy, embryo research, the ownership of stem cells,
professional role conflict and, finally, criteria for access to
NRTs. Most of these options are in a stage of rapid development with few commentators predicting that this
development will cease. As a result, reprogenetics will
continue to raise ethical challenges and to reinterpret the
meaning of responsibility and other moral concepts. The case
studies and the activities in this workbook, for example, call
for new interpretation of these concepts which include social
aspirations to promote: the 'welfare' and 'well-being' of
offspring; the reduction of human suffering; respect for
human embryonic life; the liberty of individuals to make
reproductive decisions they judge responsible; and last but,
not to be minimised, the nurturing of the 'public good'.
Discussion of cases made it clearer that human goals and
values are often in competition or tension. This tension
demonstrates clearly that moral disagreement is a common

occurrence which requires careful scrutiny and negotiation especially when it comes to prioritising or applying ideals and principles to concrete cases. But even admitting the inescapable reality of moral disagreement, decisions still have to be made.

Concerns about values at risk and values to be protected focus our attention on social goals and questions of whether or how these goals and values will be incorporated in structures of access to NRT use. The task ahead may well be the complex one of human deliberation to maximise the opportunities to decide:

1. Whether and to what extent NRTs will be provided by any given society
2. How such provisions will be managed with social responsibility in diverse societies
3. Each of these tasks assume a further question: *Who will be the decision makers?*
4. They also reflect and reinforce the most fundamental of moral questions: *How should we live?*

Let's take a moment here to read one philosopher's understanding of that social concern about *how we should live?* Looking back over many of the ethical questions raised by reprogenetics in this TEMPE text might make more visible the fact that this question *'how should we live?'* was in the background all the time. We will focus this question in this final chapter by looking at the attempts by many societies, in their efforts to 'ethically manage NRTs', to balance the values of individual liberty and the public good. Michael Parker recognises the social necessity to try and harmonise these values as he looks at the use of genetics in reproductive choice:

> The need to make decisions about public policy in this area...raises important questions about the relationship between the public and the private in ethics. When, for example, are my reproductive choices, if ever, a matter for me and my family alone, to be made in the privacy of our own home, and when are they, again if ever, a matter for public decision making? In some ways genetics, particularly when combined with new reproductive technology, seems to bring the private into the public arena. (Parker 2000: 160)

Following on Parker's point, there are a number of ways in which the availability of NRTs has brought reproductive decision making into the public arena.

- In the past, people often had children without forethought and planning – by 'chance' rather than 'choice'. Couples who reproduced were not required to have counselling to see if they would be 'fit and responsible parents'. Today, access to use of NRTs in some countries require rather careful consideration of the stability of the relationship of the couple seeking fertility assistance. These are seen as efforts to try and ensure 'responsible' parenting becomes a focus under new reproductive choices.
- Up to now, parents were not offered or encouraged to have genetic tests to see if the couple were carriers for any serious family diseases. Today, the welfare of future children is a consideration brought to the forefront with available genetic testing.
- Until recently, women could not have chorionic villus sampling (CVS) or amniocentesis or ultrasound to find out the health and progress of their future child(ren). Today, all of the options mentioned here are part of health care provisions in many countries and might be viewed by some as a part of 'responsible parenting'.

Activity:

Stop for a few moments and see if you can think of any additional ways in which the advent of NRTs has made the private, public.

The move from chance happening in the private sphere to choice decision making in the public domain means that reproduction is now greatly monitored by policy guidelines, ethical scrutiny or legal protections and sanctions for abuse of procedures. Now, if we recognise the intersection between the private choices of people wishing to have children and the public, institutional provisions or prohibitions in this regard that have been illustrated in the case studies of this text, two dominant values become explicit. These values are individual liberty and the public good and they are teased out in an extract from Dolores Dooley's article on reproductive responsibility below.

Moral Free-Fall in Ethics: Re-thinking Reproductive Responsibility

Dolores Dooley

Individual liberty

The positive endorsement of liberty for individuals is not new. Long before John Stuart Mill formulated his liberty principle in 1859 with the publication of *On Liberty,* political philosophers and theologians alike had argued the centrality of liberty to individual and social well-being. Drawing from this long established human value we might ask: Should NRTs primarily be viewed as individual choice options about fertility and family planning? If so, then decisions about their use or control would be decisions about justified limits to the exercise of human liberty. A country that values individual liberty might allow NRTs to be widely available and not subject to strict criteria for access. They might, however, be regulated by policy guidelines of law to protect the rights of clients, future offspring and clinicians in the fertility transaction.

Public good

Other countries where decision makers are more sceptical of a generous social provision of liberty options will either prohibit many forms of NRT or will circumscribe their use with fairly detailed legal requirements to try and ensure compliance of liberty choices with some overall view of what is in the 'public or common good'. The inspiration for this overall view of the public good might derive from cultural traditions to do with procreation, ethnic considerations of respect for foetal life, or perhaps religious perspectives on sexuality and reproduction.

Another pivotal concern in social questioning about NRTs is the protection of individuals who might be considered vulnerable or disadvantaged in some way when a society allows for individual liberty of access to NRTs. Some citizens or individuals who might be considered vulnerable as a result of such NRT practices are: people with impairments, the economically marginalised, women and parents. Thus, to protect all citizens, a society would look closely at NRT practices and developing technologies and ask: How can responsible use be managed so that protection of vulnerable groups is given sustained attention? A society's efforts to be responsible require that it give an accounting of the decisions it takes about NRT availability. Social responsibility requires that members of any given community be prepared to be accountable, to precisely give an accounting of the value bases for the provision or prohibition of NRTs. Within a democratically run society at least, this social accounting also needs to be made credible within the relevant conversational contexts: professional, familial, legal, educational and ecclesial. The objective, then, for a society aiming at responsible accounting is, through a range of mechanisms, to best integrate and protect the joint values of 'liberty' and 'public (social) good' (Dooley 2000).

In order to explore the ethical tensions that might arise in trying to harmonise values of liberty and the public (social) good we begin with a case study – a variation on an actual case that occurred in Germany in 1995.

The Case of Helena and Paul

Helena and Paul are both carriers of a cystic fibrosis (CF) gene. They have decided to apply to the Assisted Reproduction Commission in their country for access to IVF facilities and access to preimplantation diagnosis of the resulting fertilised embryos. Helena and Paul have already borne a child with cystic fibrosis and their love for her is not in question. But following the birth of their daughter they subsequently aborted two further pregnancies after prenatal tests showed that the same illness affected both foetuses. Both partners feel that they cannot justify having another child that would have a life of such relentless distress and suffering. They also believe that the stress of rearing another child requiring extensive care might be more than their marriage can bear. Their final justification is that it would be unreasonable to expect them to expose themselves to the risk of a further abortion. (Graumann 2000b)

Discussion

There are at least three different reasons given to support the couple's request for PGD:

1. It is unfair (unjustified) to a future child to allow CF to affect their quality of life if such could be prevented.
2. Stress of another CF affected child would, in all likelihood, be too stressful for the marriage to sustain.
3. It is unacceptable that they should expose themselves to another abortion. (Given their views in the first two reasons here, an abortion would be their option if their embryo were affected again.)

There is another reason here not as explicitly formulated by our couple. They recognise that they are both CF gene carriers and this decision for preventing another CF child is something they consider to be an exercise of 'responsible parenting'. (See Appendix 3 for information on legislation in relation to PGD in the EU.)

Activity:

Pause and try to formulate your views here. From your moral perspective, are the arguments given to avoid having another child with CF good arguments? Notice that even the process of evaluation of reasons is a process that is normative and value laden. If you think these are 'good' reasons the couple offers, then you are thereby saying something about your own moral views. This would also apply of course if you judge that the reasons are not 'good' reasons.

You might argue that this couple's choice of PGD is simply a wrong choice, an immoral choice because it involves the willingness to destroy preimplanted embryos. Equality of respect for human life in all its stages would require protecting that life. But, what if the couple does not hold your moral views on this? Perhaps the couple firmly believes it would not be respect but rather disrespect for human life if human suffering could be prevented by this decision and was not so prevented. They would argue that 'respecting human life' means at least trying to ensure, as best we can, that the newborn would start with a 'good quality of life'. Here we ask you to review the cases of Linda and Philip and Mary and Robert in Chapter 1. In the case of Linda and Philip, the couple wanted to have PGD, in part, so that they might implant an embryo carrying the gene for deafness. In the case of Mary and Robert, a couple carrying Tay-Sachs disease ignore their doctor's advice not to conceive even though there is a 25 percent likelihood that any child they produce will have the disease. Also take a few minutes to review the Nash case in Chapter 4 where a couple had a six-year old girl, Molly, who suffered from the fatal condition of Fanconi Anaemia which prevents production of bone marrow. The Nash couple wanted another child and, through IVF and PGD, bore a son, Adam, who, in turn, supplied healthy bone marrow for Molly. However, the couple chose to destroy thirteen of the embryos from IVF which were affected with the disease that Molly had or which were not tissue compatible. Compare these cases with the case of Helena and Paul. Each of these couples deeply believed they were 'respecting human life' in their decisions. They likewise argued that their decision was made with a view to their 'responsibility' as parents. Fundamental differences of interpretation and application of 'respect for human life' are at issue.

Activity:

If the requests cited in our cases would be facilitated in your country, on what value basis are they offered? Is it the belief that, in relation to reproductive decisions, individuals should have the liberty to decide what is in their own best interests and the interests of their offspring? If the request would not be facilitated in your country, try to formulate the reasons why. Are they straightforwardly reasons to do with the belief that liberty is not a primary good for individuals? Or, perhaps the explanation for nonavailability has to do with complex cultural, political and religious reasons. Try to formulate reasons for various countries' positions that might help clarify bases for moral difference and ways of adjudicating these differences.

One perspective on the dynamics between liberty and society is given by Sigrid Graumann (2000b) who believes that reprogenetic decisions are reflections of a society's character. For her, they also shape the quality of a society's life. Graumann offers a socioethical framework for thinking about cases where disease, disability and difficult decisions are at issue. Her framework makes visible one interpretation of individual liberty decisions impacting on a society. Consider Paul and Helena's decision and the case of Molly and Adam Nash from Chapter 4 when you read the following extract.

Prenatal selection and eugenics: on the possible social consequences of reprogenetics

Sigrid Graumann

To understand arguments against prenatal and preimplantation genetic diagnosis, we will examine the notion of 'social reality' and make explicit the dynamics of change from the interaction between social reality and individuals. Social reality is a 'matrix' or 'field' in which individual persons gain their knowledge of reality, construct their convictions and opinions and make their decisions and act. That means that single autonomous decisions are moulded by social values, norms and role expectations. At the same time, the convictions, attitudes, opinions, decisions and acts of the individuals affect social reality in return (Berger and Luckmann 1999.) As a result, the individual human being participates in changing and developing social reality. This process of changing social reality is largely unconscious and involuntary but, nevertheless, intentionally influenced by single individuals. By bringing the transformation of social values and norms to the level of consciousness, options for acting are opened up which make it possible to shape social reality in a responsible fashion. And precisely that is what supporters of various positions call for when looking at a sociocultural understanding of reprogenetics.

While the transformation of social values has begun with the advent of prenatal diagnosis, it takes on a new quality through the introduction of PGD. This transformation is visible in the increasing pressure towards perfection to which potential parents feel themselves exposed. Further, it leads to an increasing stigmatising of and discrimination against those with chronic illnesses and the disabled.

1. Pressure on parents

The fact that more options are available does not necessarily mean that one is more free to act. Experience with prenatal diagnosis shows that individual decisions can express internalised social ascriptions of value to human life or even be made under direct pressure from one's family or peers. From a socioethical perspective, then, developments must be rejected which limit the possibility of making autonomous and responsible decisions – especially when they are concerned with serious moral problems. So, this would be a first socioethical argument against prenatal and PGD diagnosis.

undermines their ability to act responsibly: their increased choice limits their freedom to choose autonomously.

What is your response to her conclusion? Consider the opposite argument: that a couple's freedom is limited or harmed by banning PGD.

Now consider her second argument below which claims that allowing couples the choice requested here is to encourage a discriminatory social attitude towards the already living people with disabilities.

2. Discrimination of people with disabilities

The further social consequence of PGD would be the discrimination of the disabled. It is not being argued that the individual decision of a couple or a woman to make use of prenatal diagnosis or PGD is a discriminatory act *per se*. It is rather the transformation of social values itself as it interacts with many individual decisions, social expectations and the ascriptions of value to human life that they may entail. These multiple decisions can further the extent to which our already antidisabled society becomes more so. It is to be feared that this will lead to the social legitimation of an increasing discrimination, stigmatising and decreasing solidarity with the chronically ill, the disabled and their families. And this is the second socioethical argument against prenatal and preimplantation genetic diagnosis.

Discussion

Before we can assume that the outcome of continuing use of genetic testing for reproduction forecast by Graumann is accurate, what kind of research needs to be done? We might consider undertaking empirical studies on public attitudes and psychosocial changes towards people with disabilities if we are going to avoid decisions based solely on speculation. We could also try to ascertain, in some systematic empirical way, the attitudes of concerned professionals (reproduction doctors, human geneticists, counsellors, psychologists and others) toward PGD and prenatal diagnosis.

There is always a danger that our fears may stop developments and liberty options especially if the fears are inchoate
or without some foundation in research. Consider the following position on the necessary reliance of ethics on empirical
research:

> Empirical and prognostic studies that would help to answer questions
> in ethical discussions of PGD are available in very few areas. It is,
> therefore, likely that these questions will either be ignored, which
> would lead to an impermissibly narrowed perspective, or that many
> aspects of the discussion will be based on speculation. Empirical
> studies...belong indirectly to ethical reflection, because they are
> essential to the ethical evaluation. (Mieth et al. 2000: 330)

Given the position of Mieth et al., it could be argued that it is
important to make distinctions between different uses of PGD
so that one does not reject legitimate uses at the same time as
rejecting those that, maybe, should be deplored.

PGD for enhancement or therapeutic purposes?

A common distinction made when discussing the use of PGD
is that between genetic screening for 'therapeutic' purposes or
for 'enhancement' purposes.

Enhancement use of PGD: the use of genetic techniques to
improve genetically related traits that, without improvement, would
lie within normal ranges (Mason and Tomlinson 2001: 180).

The broad term often used to describe the goal of enhancing
the 'normal' human condition is **'eugenics'** which in turn is
distinguished between positive and negative eugenics.

Positive eugenics utilises the results of genetic research to try and
produce people with 'superior' phenotype profiles.

'Phenotype' refers to the observable biochemical or physical
features of an organism caused by both genetic make-up and envi-
ronmental causes. Some examples are skin pigmentation, eye
color, body build, etc., (Mason and Tomlinson 2001: 180).

> *Negative eugenics* tries to improve the human condition by elim-
> inating biologically 'inferior' people or unwanted clusters of
> phenotypes from the population (Mason and Tomlinson 2001: 180).

It is clear that both forms of enhancement presuppose crucial value and scientific judgments about 'normal' and 'abnormal', desirable and undesirable, acceptable and unacceptable forms of human beings.

While there has been a great deal of debate surrounding the use of PGD for enhancement or eugenic purposes, many, who disagree profoundly with using genetics for these goals, nevertheless, argue that the '*therapeutic use of PGD*' is a legitimate use of human freedom. Therapeutic PGD allows people willing to take on the difficulties of IVF treatment to have the option of selecting only 'healthy' embryos in their efforts to have children. Embryos which have a gene predisposing them to a particular condition can be discarded before implantation allowing for the *in utero* development of only the embryos which are free from the so called 'disease' gene. This is, partly, what happened, for example, in the Nash case discussed in Chapter 4.

Yet Graumann's discussion offers a cautionary note to be vigilant of the extent of social construction in the very determination of 'disease'. Michael Reiss explains this notion of social construction in the following example:

> A disease is, in a sense, a relationship a person has with society. Is being 4 ft. tall a disease? The answer tells us more about a society than it does about an individual of this height. Some conditions are relatively unproblematic in their definition as a disease. For instance, Lesch-Nyhan disease is characterised by severe mental retardation, uncontrolled movement (spasticity) and self-mutilation. No cure is at present available and the person dies early in life, after what most people would consider an unpleasant existence... (Reiss 2001: 25)

Moreover, there is an emerging disability rights movement built on the shared belief that many problems experienced by persons with disabilities, problems which seriously interfere with the quality of their lives are caused, not solely or even most importantly by the impairment, but by the ignorance of other people, the fear of difference and 'by the barriers that exist in society, whether they are architectural, institutional, technological, legal or attitudinal' (Kaplan 1999: 131).

Thus, Graumann's concerns about the possible stigmatisation of people with disabilities as a result of the availability of therapeutic PGD are legitimate social concerns. The normative judgments, such as those outlined above, about what are 'diseases' and 'disabilities' are decisive judgments in making choices about the ends for which PGD will be used. That they are decisive was acutely demonstrated in our discussions following the deaf case study discussed in Chapter 1.

On the other hand, in the passage above, Reiss claims that describing Lesch-Nyhan disease as a disease is unproblematic because its symptoms are so debilitating. It is, in the words of John Harris, outlined in Chapter 1, a 'harmed condition'. Accepting that some forms of disability are inherently undesirable, Harris compares gene therapy with other therapeutic procedures in medical science which attempt to cure disability.

> On the view assumed here...all persons share the same moral status whether disabled or not. To decide not to keep a disabled neonate alive no more constitutes an attack on the disabled than does curing disability. To set the badly broken legs of an unconscious casualty who cannot consent does not constitute an attack on those confined to wheelchairs. To prefer to remove disability where we can is not to prefer non-disabled individuals as persons... In so far as gene therapy might be used to delete specific genetic disorders in individuals or repair damage that had occurred genetically or in any other way, it seems straightforwardly analogous to any other sort of therapy and to fail to use it would be deliberately to harm those individuals whom its use would protect. (Harris 1999: 167)

If we accept the distinction Harris and Reiss make between 'disabilities' that are, largely, socially constructed and disabilities that are, largely, 'harmed conditions' and apply Harris' discussion of gene therapy to the case of PGD, two responses to Graumann's concerns are possible. First, PGD could, justifiably, be used therapeutically: in order to delete or repair genetic disorders of a serious kind. Second, because society casts impairments in certain ways thereby contributing to our conception of them as disabilities, the barriers to the equal treatment of people with impairments, listed by Kaplan above, could be viewed as a list of societal goals – architectural improvements, institutional acknowledgement and inclusion, technological progress, legislative supports and education. In this way, both individuals and society may benefit from the use of PGD for therapeutic purposes while, at the same time, society as a whole becomes more proactive in the way in which it addresses the sociocultural and institutional discrimination of people with impairments.

Activity:

When we make a judgment that genetic testing for purposes of reproduction will inevitably cause discrimination of the disabled and chronically diseased, consider whether or not we are basing our judgment on empirical evidence, or are the judgments coming from fears and worries based on historical and literary caricatures of disabled people as 'negative' beings?

Sex selection and inequality?

Francoise Shenfield works in fertility medicine in London and recognises that genetic testing in reproduction has rekindled the debate about sex selection for social reasons, for instance, when several 'healthy' embryos of different gender potential would be available for embryo transfer. Where serious diseases are sex-linked many fertilisation centres are positive about providing PGD to disease affected embryos and in doing that they are only coincidentally choosing to omit a particular sexed embryo. That would be sex selection for medical/therapeutic reasons.

But Shenfield agrees with the view that sex selection is likely to reinforce sexist attitudes already too prevalent in most societies. She admits that this should be a most powerful argument against whimsical sex selection. But she sees the burden of another kind of sex-linked inequality, that of female ageing and decreased ability to procreate which may be relieved by PGD. Here Shenfield argues that the use of PGD can enchance the liberty choices of older women who, for any number of complex reasons did not reproduce earlier in their child-bearing years. With PGD these women would experience pregnancy with greater peace of mind and confidence in a 'successful' pregnancy (Shenfield 2000).

The Health Council of the Netherlands provides a researched study on 'Sex Selection for Non-Medical Reasons' which allows for empirical data to be factored in to our concerns that inequality and discrimination could result from the use of PGD. The report deserves a close study by anyone concerned about sex selection eventualities under reprogenetics. In a final paragraph the Council writes:

> The possibility cannot be excluded that the actual availability of sex selection insemination as a family-planning instrument will result in parents finding the sex of their children more important than they

claim at present. Because choice involves making distinctions, the possibility of sex selection could result in the undermining of the idea of sexual equality and therefore of the struggle for emancipation. Making sex an object of choice could lead to the reinforcement of stereotypical ideas about sexual roles. (Health Council of the Netherlands 1995: 37)

The Council of the Netherlands offers a paradigm of the need to combine the best of empirical research with clear focus on normative questions to do with inequality and potential social discrimination that might arise with reprogenetic uses.

Activity:

(a) The Council of the Netherlands claims above that '[m]aking sex an object of choice could lead to the reinforcement of stereotypical ideas about sexual roles'. In the light of our discussion regarding Graumann's concerns with PGD, do you think that the claim would have the same force if it were applied to disease or disability? Rephrased, the claim looks like the following:

'Making disease or disability an object of choice could lead to the reinforcement of stereotypical ideas about disease and disability.'

(b) Given the rapid pace of development of NRTs, it might be argued that calls for caution, reflection and responsibility are straws in the wind of change. Indeed, it is unlikely, given the interests at stake in these technologies – personal, social, corporate – that these developments will be halted or reversed. Granted the current situation, how do you think a society such as your own should respond?

How should we live?

This chapter started with the question considered central to this TEMPE text: 'How should we live?' In continuing a reflection on this question we inevitably confront the challenge of social responsibilities to integrate and harmonise liberty and the public good. One such ongoing attempt to do this comes from the Human Fertilisation and Embryology Authority (HFEA) which regularly has to stop and consider its legitimacy, that is, consider how it reaches decisions and the ethical basis on which it does so. In its decision making, the Authority tries to find certain principles that underpin HFEA Code. Four main principles have recently been reaffirmed with the added proviso that 'a regular feature of any system should be consultation' – with all the 'stake holders' in the development of NRTs. These principles for decision making from the HFEA could form a solid basis for discussion as we end this examination of the ethics of NRTs.

- The respect which is due to human life at all stages in its development;
- The right of people who are or may be infertile to the proper consideration of their request for treatment;
- A concern for the welfare of children, which cannot always be adequately protected by concern for the interests of the adults involved; and,
- A recognition of the benefits, both to individuals and to society, which can flow from the responsible pursuit of medical and scientific knowledge. (Whittal 2000)

Activity:

Examine each of the HFEA principles in turn and place them in an order of importance. Consider the practical implications that your ordering might have for the availability and development of NRTs.

In what follows we have compiled a table of some of the arguments for and against PGD that have been presented in this chapter. Take a few minutes to read the table and see if you can construct more arguments of your own.

Arguments for and against PGD

For	Against
✔ respecting human life means trying to ensure that the newborn starts with a good quality of life	✘ PGD involves a willingness to destroy preimplanted embryos

✔ banning PGD limits the choice of couples to ensure, insofar as they can, that their future children will be free of genetic disease

✔ it is like other therapeutic procedures in medical science and failing to use it deliberately harms those individuals who should be protected

✔ it enhances the freedom of choice of older women who can contemplate childbearing with greater confidence in a successful pregnancy

✔ ?

✘ increased choice puts psychological and social pressure on parents which limits their freedom to choose autonomously

✘ it legitimates the discrimination and stigmatisation of people with disabilities

✘ sex selection for medical/therapeutic reasons which may follow on PGD is likely to reinforce sexist attitudes

✘ ?

Activity:

Which position do you find most convincing and on what grounds?

Summary

This Chapter focused on the notion of social responsibility and, specifically, on the relationship between individual liberty and the public good.

We pointed out that

• the pace of reproductive technological development calls for a rethinking of fundamental ethical concepts and a re-evaluation of traditional assumptions and positions

• tensions arise between the values of individual freedom and social concerns in relation to reprogenetics

• one of the challenges with regard to some reprogenetic technologies, such as PGD, is to ensure that genetic testing does not promote discrimination on disability or gender grounds

Appendices

Appendix 1

EU Eligibility Criteria for Access to Assisted Reproduction

Country	Legislation	Regulatory or Legislative Criteria
Austria	Act No. 275, 1992	Married or stable heterosexual cohabiting couple who are able to provide a satisfactory home and have tried all other fertility treatments.
Belgium	None	No legislative criteria (NLC). In practice, AR is available for single and lesbian women at more than five Flemish centres.
Denmark	Law No. 499, 1996; No. 460, 1997	NLC. In practice, treatment is provided only to alleviate infertility and is restricted to married women, under 37 years.
Finland	Proposed	Proposed legislation restricts access to married or heterosexual cohabiting couples who are involuntarily childless or at risk of transmitting a serious disease and where the woman is under 50 years.

France	Law 94–654, 1994	Treatment is provided only to alleviate infertility or to avoid the transmission of serious disease. It is limited to married or cohabiting heterosexual couples of reproductive age. Posthumous insemination is prohibited.
Germany	Embryo Protection Act, 1996	NLC. In practice, treatment is typically provided only to married couples though specific committees have the power to consider the case of single persons.
Greece	None	NLC. In practice, treatment is provided only to married or cohabiting heterosexual couples.
Ireland	None	NLC. In practice, AR is provided under the Guidelines of the Medical Council and is typically restricted to married couples.
Italy	None	NLC. In practice, AR is provided under the Code of Medical Deontology which restricts treatment to stable heterosexual couples for the alleviation of infertility and prohibits posthumous insemination, any form of surrogacy and the treatment of elderly women. Some Italian clinics, however, were among the first to treat postmenopausal women.
Luxembourg	None	NLC. In practice, no AR treatments are available.
Netherlands	Hospitals Act, Decree on IVF Planning 1989	NLC. In practice most IVF centres limit access to women under 40.

Portugal	None but proposed	NLC. In practice, treatment is offered to stable heterosexual couples.
Spain	Unimplemented Law 35, 1988	NLC. Every woman is eligible for treatment as long as she is over 18, mentally competent and provides written consent.
Sweden	Law 711, 1988; Law 1140,1984	Treatment is only provided to alleviate infertility and is restricted to married or cohabiting heterosexual couples.
UK	HFE Act 1990	The welfare of the child (including the need for a father) must be taken into account. In practice, clinics consider age, duration of infertility and likelihood of success.
European Convention on Human Rights and Biomedicine 1996	So far, the 10 EU signatories are: Denmark, Finland, France, Greece, Italy, Luxembourg, Netherlands, Portugal, Spain and Sweden.	'Parties to this Convention shall protect the dignity and identity of all human beings and guarantee everyone, without discrimination, respect for their integrity and other rights and fundamental freedoms with regard to the application of biology and medicine.' Article 1. 'The interests and welfare of the individual human being shall prevail over the sole interest of society or science.' Article 2. 'Parties, taking into account health needs and available resources, shall take appropriate measures with a view to providing, within their jurisdiction, equitable access to health care of appropriate quality.' Article 3.

(Sources: Council of Europe 1996; Beyleveld and Pattinson 2000; EGE 2000; Lee and Morgan 2001)

Appendix 2

EU Legislation in Relation to Embryo Research, Germ Line Gene Therapy and Reproductive Cloning

Country	Embryo Research	Germ line Gene Therapy	Therapeutic and Reproductive Cloning
Austria	Act. No. 275, 1992. Prohibited unless therapeutic (nondestructive).	Act No. 275, 1992. Prohibited.	Act No. 275, 1992. Prohibited.
Belgium	Proposed legislation. Permitted by default, subject to conditions, e.g., up to 14 days.	Proposed legislation.	Proposed legislation.
Denmark	Law No. 460, 1997, ss25-28. Permitted, subject to conditions, e.g., up to 14 days.	Law No. 460, 1997. Prohibited.	Law No. 499, 1996; Law No. 460, 1997. Prohibited.
Finland	Act No. 488, 1999. Permitted, subject to conditions, e.g., up to 14 days.	Act No. 488, 1999. Permitted where it aims to cure or prevent serious disease.	Act No. 488, 1999. Prohibited.
France	Law 94–654, 1994; Decree 97–578, 1997. Permitted only if it doesn't impair the embryo.	Law 94–654, 1994. Prohibited.	Law 94–653 and 94–654, 1994. Prohibited.

Germany	Embryo Protection Act, 1990. Prohibited unless therapeutic.	Embryo Protection Act, 1990. Prohibited.	Embryo Protection Act, 1990. Prohibited.
Greece	No legislation. Permitted by default, subject to conditions, e.g., up to 14 days	No legislation.	No legislation. Prohibited by regulation.
Ireland	8th Constitutional Amendment Implicitly prohibited.	No legislation. Implicitly prohibited.	No legislation. The legal position is uncertain.
Italy	Proposed legislation. Therapeutic research permitted by default.	No legislation.	No legislation.
Luxembourg	No legislation, no research.	No legislation.	No legislation.
Netherlands	Proposed Embryo Bill. Permitted by default, subject to conditions, e.g., up to 14 days.	Proposed legislation. Recommends a 5-year moratorium.	Proposed legislation will delay therapeutic cloning and prohibit reproductive cloning.
Portugal	No legislation. Permitted by default.	No legislation.	No legislation.
Spain	Unimplemented legislation which permits research subject to conditions, e.g., up to 14 days.	Unimplemented legislation. Prohibits research where the non-pathological genetic patrimony is modified.	Law 35, 1998; Title V of Penal Code. Prohibited if aimed at race selection.

Sweden	Law No. 115, 1991. Permitted subject to conditions, e.g., up to 14 days.	Law No. 115, 1991. Prohibited.	Law No. 115, 1991. Prohibited.
UK	HFE Act 1990. Permitted subject to licencing requirements and conditions, e.g., up to 14 days.	HFE Act 1990. Permitted only by regulation. (No regulations currently exist.)	HFE Act 1990. Therapeutic (not reproductive) cloning (up to 14 days) is permitted.
European Convention on Human Rights and Biomedicine 1996	1. Where the law allows research on embryos in vitro, it shall ensure adequate protection of the embryo. 2. 'The creation of human embryos for research purposes is prohibited.' Article 18.	'An intervention seeking to modify the human genome may only be undertaken for preventive, diagnostic or therapeutic purposes and only if its aim is not to introduce any modification in the genome of any descendants.' Article 13.	'Any intervention seeking to create a human being genetically identical to another human being, whether living or dead, is prohibited.' Protocol on the prohibition of cloning human beings.

(Sources: Council of Europe 1996; EGE 2000; Beyleveld and Pattinson 2000; Lee and Morgan 2001)

Appendix 3

EU Legislation in Relation to Abortion, Prenatal Diagnosis and Preimplantation Genetic Diagnosis

Country	Abortion	Abortion following Prenatal Diagnosis	Preimplantation Genetic Diagnosis
Austria	S. 97 of Penal Code. Permitted with conditions.	Allowed up to birth subject to conditions.	Act No. 275, 1992. Implicitly forbidden.
Belgium	Law, April 1990. Permitted with conditions.	Allowed up to birth subject to conditions.	No legislation. Permitted by default.
Denmark	Pregnancy Act 1973; No. 499, 1996. Permitted on demand up to 12 weeks.	Allowed up to birth subject to conditions.	Law No. 460, 1997. Implicitly permitted.
Finland	Title of legislation unknown. Permitted with conditions.	Allowed up to birth subject to conditions.	No legislation. Permitted by default.
France	Law 94–654, 1994; Decree No. 95-559, 1995; Decrees 97–578 and 97–579, 1997. Permitted on demand up to 10 weeks.	Allowed up to birth subject to conditions.	Law 94–654, 1994. Currently prohibited. Unimplemented law allows PGD with conditions, e.g., risk of incurable genetic defect.

Germany	Criminal Code s.218. Abortion is a criminal offence but neither the physician nor the woman is prosecuted in specified circumstances.	Allowed up to birth subject to conditions.	Embryo Protection Act 1990. Implicitly prohibited.
Greece	Law 1609, 1986; No.1036, 1980. Permitted with conditions.	Allowed up to 24 weeks subject to conditions.	No legislation. Permitted by default.
Ireland	8th Amendment to the Constitution. Prohibited unless there is a real and substantial threat to the life of the mother.	Limited provision of PND.	8th Amendment to the Constitution. Implicitly prohibited.
Italy	Legge No.194/78, 1978. Permitted on demand up to 80 days.	Allowed up to birth subject to conditions.	No legislation. Permitted by default.
Luxembourg	Title of legislation unknown. Permitted with conditions.	Allowed up to 12 weeks with conditions.	No information. No information.
Netherlands	The Pregnancy Termination Act, 1981. Permitted up to 24 weeks with conditions.	Allowed up to 24 weeks with conditions.	No Legislation. Permitted by default.
Portugal	Arts. 140, 141, 142 Penal Code. Permitted with conditions	Allowed up to 24 weeks with conditions.	No Legislation. No information.

Spain	Art 417, 1985, Decree No. 2409, 1986. Permitted with conditions	Allowed up to 22 weeks with conditions.	Unimplemented legislation. Permitted with conditions.
Sweden	Abortion Act, 1995. Permitted on demand up to 18 weeks.	Allowed up to 22 weeks with conditions.	Law 115, 1991. Permitted with conditions, i.e., for diagnosis of serious, incurable hereditary disease.
UK	Abortion Act 1967; HFE Act 1990. Permitted on demand up to 24 weeks and thereafter with conditions	Allowed up to birth with conditions.	HFE Act 1990. Permitted with conditions, i.e., licence and diagnosis of chromosomal abnormalities.
European Convention on Human Rights and Biomedicine 1996	'Parties to this Convention shall protect the dignity and identity of all human beings ...' Article 1.		'The use of techniques of medically assisted procreation shall not be allowed for the purpose of choosing the future child's sex, except where serious hereditary sex-related disease is to be avoided.' Article 14.

(Sources: Council of Europe 1996; Beyleveld and Pattinson 2000; EGE 2000; Lee and Morgan 2001)

Glossary

azoospermia: the absence of spermatozoa in the semen or failure of formation of spermatozoa

cell line: is a group of stem cells* that is capable of sustaining continuous, long-term growth in a laboratory culture.

clone: a collection of genetically identical cells or organisms.

chorionic villus sampling (CVS): this is an antenatal screening test to determine the condition of the foetus. CVS is usually carried out between 8 and 11 weeks of pregnancy. It can be performed vaginally, through the uterine cervix, or abdominally, involving the extraction of a sample of the chorionic villi (the tree-like projections of trophoblast which burrow into the decidue to form the placenta. This test is used to detect:

- Chromosomal abnormalities such as Down's syndrome
- To enable DNA analysis (for cystic fibrosis, etc.)
- Inborn errors of metabolism
- The sex (where sex-linked diseases are in question)

consequentialist: see utilitarian*.

cryopreservation: the freezing of sperm, ova, embryos or any body parts and tissues.

deontological: (From Greek '*deon*', what is due) describes moral systems which judge actions according to their intrinsic (noninstrumental) merit or demerit and which emphasises duties and the rights to which they give rise. The most influential deontological system is Kantianism, which stresses autonomy and respect for persons.

embryo: the developing organism between conception and eight weeks after which it is referred to as a foetus up to the time of birth. During the first fourteen days of an embryo's development the name pre-embryo is sometimes used. (See embryonic development below.)

embryonic development: until implantation in the uterine wall, the embryo passes through the following five stages of development.

1. The *zygote* stage: at two to three days after fertilisation, the embryo consists of *totipotent** identical cells each of which could give rise to an embryo on its own. These cells are totally unspecialised and have the capacity to differentiate into any of the cells which constitute the foetus and what surrounds it.

2. The *morula* stage: at four to five days after fertilisation, the number of cells increases rapidly to 12 – 16 cells. These *pluripotent** cells are still unspecialised but they can no longer give rise to an embryo on their own.

3. The *blastocyst stage*: at five to seven days after fertilisation a hollow appears in the center of the morula and the cells constituting the embryo start to differentiate into inner cells which give rise to the foetus and surrounding tissue and outer cells which constitute the placenta and other surrounding tissue.

4. *Implantation* can follow once the embryo has grown out of the oolemma (hatching). Implantation begins on approximately the 7th day after fertilisation and is only completed around the 14th day. In this process, the outer cells (trophoblast) penetrate the mucous membrane of the uterine wall, to assist in the formation of the placenta.

5. The *primitive streak*, a groove that appears in the embryonic disc about 14 – 15 days after fertilisation. It is taken to be the first sign that an embryo will develop; if the groove does not form, embryonic development does not progress.

eugenics: positive eugenics utilises the results of genetic research to try and produce people with 'superior' biochemical or physical features such as skin pigmentation, eye color, body-build, sexual orientation, etc. Negative eugenics tries to improve the human condition by eliminating biologically 'inferior' people with unwanted biochemical or physical features.

germ cell: there are two classes of cells found in the human body: germ cells and somatic* cells. Germ cells are found in the ovaries of a female and the testes of a male. They give rise to ova and sperm respectively. Somatic cells are all the other cells in the body.

insemination: the introduction or injection of semen into the reproductive tract of (a female)

IVF (in vitro fertilisation): is a technique which involves the fertilisation of ova outside of a woman's body. *In vitro* literally means 'in glass' as opposed to *in vivo* meaning 'in a living thing or organism'. In the usual form of IVF, sperm are added to laboratory dishes or test tubes containing eggs. Some or all of the fertilised egg(s) are later transferred to a woman's uterus.

nuclear family: a two-parent heterosexual family.

oocyte: an immature ovum in an ovary.

pro-nucleus: a small round structure seen within the egg after fertilisation which contains the genetic material of each gamete surrounded by a membrane. A normal fertilised egg should contain 2 pro-nuclei, one from the egg and one from the sperm.

reproductive cloning: would occur if the embryo created through somatic cell nuclear transfer* were successfully implanted in a womb and developed to full term. The baby, subsequently born, would be genetically identical with the individual from whose somatic cell he or she originated from.

stem cells: these cells have different forms with varying abilities to differentiate into specialised tissues.

- *Pluripotent stem cells* are derived from the blastocyst formed in early embryonic development* and may give rise to many cells in an adult animal. They have an extensive capacity to divide and the potential to develop into most of the specialised cells or tissues in the body. When they are derived from an embryo they are termed embryonic stem (ES) cells. When they are derived from germ cells in an embryo (from the tissue that is destined to develop into the sperm or ova) they are termed embryonic germ (EG) cells*.

- Some authors equate *pluripotent* and *totipotent* cells, but others distinguish between totipotent cells, at the 2–4 cell stage, which are capable of giving rise to every cell line in the body and to an entire human individual; and pluripotent cells, which are derived from the blastocyst at a slightly later stage of development, when the outer and internal cells have already become differentiated.
- *Multipotent* stem cells can be multiplied and maintained in culture but do not have an unlimited capacity for differentiation and renewal. In the foetus, these stem cells are essential to the formation of tissue and organs. Though less abundant in the adult, they replenish tissues whose cells have a limited life span, e.g., skin stem cells.

somatic cell: see germ cell.

somatic cell nuclear transfer (SCNT): this procedure involves the creation of an embryo by inserting a somatic cell of a patient's own body into a donated human, or even animal, unfertilised egg from which the nucleus has been removed. If these reconstructed eggs are stimulated to develop to the blastocyst stage, pluripotent stem cells that are genetically identical to the patient can be derived from them. SCNT is heralded as the most successful means of avoiding immunological problems after transplantation. It is sometimes defined as 'therapeutic cloning'*. This is because related technology could lead to the cloning of human individuals if the reconstructed embryos were transferred to a woman's uterus. However, this is contrary to European Community Law and expressly prohibited in most countries.

therapeutic cloning: see somatic cell nuclear transfer

utilitarian: pertains to a consequentialist moral theory which judges actions according to their outcome or consequences and which states that the morally best outcomes are those that maximise the happiness or interests of all concerned with the action.

xenotransplantation: involves the transplantation of tissues and bodily parts from nonhuman animals to humans.

References

American College of Medical Genetics (1995). Guidelines. *American Journal of Human Genetics*, **57**: 1499–1500.

American Society for Reproductive Medicine (ASRM) (1997). Informed consent and the use of gametes and embryos for research. *Fertility and Sterility*, **68**: 780-1.

Anderson D. J. (1999). Assisted reproduction for couples infected with the human immunodeficiency virus type 1. *Fertility and Sterility*, **72**(4): 592–4.

Anderson Elizabeth (1993). *Value in Ethics and Economics.* Cambridge Mass.: Harvard.

Andre Judith, Fleck Leonard M. and Tomlinson Tom (2000). On being genetically 'irresponsible'. *Kennedy Institute of Ethics Journal*, **10**(2): 129–146.

Andrews L. (1989). *Between Strangers: Surrogate Mothers, Expectant Fathers and Brave New Babies.* New York: Harper and Row.

Baby M. 217 N.J. Supr. 313 (1987). Affirmed in part and reversed in part: 109 N.J. 396. 537A. 2d 1227 (1988).

Baran A. and Pannor R. (1993). *Lethal Secrets* (2nd edition). New York: Amistad.

Barrett Michèle and McIntosh Mary (1991). *The Anti-Social Family.* London: Verso.

Bartlett K. (1984). Rethinking parenthood as an exclusive status: the need for legal alternatives when the premise of the nuclear family has failed. *Virginia Law Review.* **70**: 879.

Bateman N. S. (1998). The medical management of donor insemination. In *Donor Insemination: International Social Science Perspectives*, K. Daniels and E. Haimes, (eds), Cambridge: Cambridge University Press, pp. 105-130.

Baudouin Jean-Louis (1999). Personal communication.

Bayles M.D. (1988). Trust and the professional-client relationship. In *Professional Ideals*, A. Flores, (ed.), Belmont CA: Wadsworth, pp. 66–80.

Beauchamp Tom and Childress James F. (2001). *Principles of Biomedical Ethics* (5th edition). New York and Oxford: Oxford University Press.

Beyleveld Deryck and Pattinson Shaun (2000). Legal regulation of assisted procreation. genetic diagnosis and gene therapy. In *The Ethics of Genetics in Human Procreation*, Hille Haker and Deryck Beyleveld, (eds), Aldershot: Ashgate, pp.215-276.

Berger Peter L. and Luckmann Thomas (1999). *Die gesellschaftliche Konstruktion der Wirklichkeit. Eine Theorie der Wissensoziologie*. Frankfurt: Fischer Publications.

Berghmans Ron, de Wert Guido and Boer Gerard (2000). Ethical issues in the use of human embryonic and foetal tissue for transplantation. Unpublished Paper, *Reproductive Ethics Workshop of the TEMPE Project*. Cork, 16–17 June.

Bernard A. and Fuller B. J. (1996). Cryopreservation of human oocytes: a review of current problems and perspectives. *Human Reproduction Update*, **2**: 193–207.

Bernat E. and Vranes E. (1996). The Austrian Act on procreative medicine: scope. impacts and inconsistencies. In *Creating the Child: the Ethics, Law and Practice of Assisted Reproduction,* Donald Evans, (ed.), The Hague: Martinus Nijhoff Publishers.

Birnbacher. Dieter (2001). Teaching clinical medical ethics. In *The Cambridge Medical Ethics Workbook*, Michael Parker and Donna Dickenson, (eds), Cambridge: Cambridge University Press.

Blum L. A. (1990). Vocation, friendship and community: limitations of the personal-impersonal framework. In *Identity, Character and Morality*, Owen Flanagan and Amalie Oksenberg Rorty, (eds), Cambridge MA: MIT Press, pp.173–97.

Blustein. Jeffrey (1982). *Parents and Children*. New York and Oxford: Oxford University Press.

Blyth E. (1994). 'I wanted to be interesting. I wanted to be able to say I've done something interesting with my life': Interviews with surrogate mothers in Britain. *Journal of Reproductive and Infant Psychology*, **12**: 189–198.

Blyth E. (1995). Not a Primrose Path: commissioning parents experiences of surrogacy arrangements in Britain. *Journal of Reproductive and Infant Psychology*, **13**: 185-196.

Boer G. J. (1994). Ethical guidelines for the use of human embryonic or fetal tissue for experimental and clinical neurotransplantation and research. *Journal of Neurology*, **242**: 1–13.

Brazier Committee Review (1998). *Surrogacy: Review for Health Ministers of Current Arrangements for Payments and Regulation* (Cm. 4068). London: Department of Health.

Brewaeys A., Ponjaert I, van Hall E. and Golombok S. (1997). Donor insemination: child development and family functioning in lesbian mother families. *Human Reproduction,* **12** (6): 1349–1359. (European Study)

British Medical Association (BMA) (1995). *Changing Conceptions of Motherhood: A Report on Surrogacy.* London: BMJ Publishing Group.

Brodzinsky D. M., Smith D. W. and Brodzinsky A. B. (1998). Children's adjustment to adoption. *Developmental and Clinical Issues,* **38**, London: Sage Publications.

Caplan G. (1968). Clinical observations on the emotional life of children in the communal settlements in Israel. In *Psychopathology*, Charles F. Reed, (ed.), Buenos Aires: Averez.

Capron A. (1999). Good intentions. *Hastings Center Report,* **29**(2): 26–27.

Chadwick Ruth (ed.) (2001). *The Concise Encyclopedia of the Ethics of New Technologies.* London: Academic Press.

Chan R., Raboy B. and Patterson C. (1998). Psychosocial adjustment among children conceived via donor insemination by lesbian and heterosexual mothers. *Child Development,* **69** (2): 443–457. (United States Study)

Childlessness Overcome Through Surrogacy (COTS) (1997). *Information for Surrogates.* London: COTS.

Cohen J., Scott R., Alikani M., Schimmel T., Munne S., Levron J., Wu L., Brenner C., Warner C. and Willadsen S. (1998). Ooplasmic transfer in mature human oocytes. *Molecular Human Reproduction,* **4**: 269–80.

Comité Consultatif National déthique pour les sciences de la vie et de la santé (1993). Avis relatif aux recherches et utilisation des embryons humains in vitro à des fins médicales et scientifiques. In *CCNE*, Dixième Anniversaire, 1983 – 1993, Paris: CCNE, pp. 113–61.

Committee on Xenograft Transplantation (1996). *Xenotransplantation: Science, Ethics and Public Policy.* Institute of Medicine.

Cooke I. (1991). *My Story.* England: Jessop Hospital for Women.

Cook R. and Golombok S. (1995). A survey of semen donors: phase II – the view of the donors. *Human Reproduction*, **10**: 951–959.

Cook R., Golombok S., Bish A. and Murray C. (1995). Disclosure of donor insemination: parental attitudes. *American Journal of Orthopsychiatry*, **65**: 549–559.

Council of Europe (1989). *Report on Human Artificial Procreation.* Strasbourg: Texts of the Council of Europe.

Council of Europe (1996). *European Convention for Protection of Human Rights and Dignity of the Human Being with Regard to the Application of Biology and Biomedicine: European Convention on Human Rights and Biomedicine.* Strasbourg: Texts of the Council of Europe.

Council For International Organisations of Medical Sciences (CIOMS)/World Health Organisation (WHO) (1993). *International Ethical Guidelines for Biomedical Research Involving Human Subjects.* Geneva: CIOMS/WHO.

de Wert Guido and Berghmans Ron (2000a). Human embryonic stem cells: ethics and policy. Unpublished Paper, *Reproductive Ethics Workshop of the TEMPE Project*, Athens. 20-21 October.

de Wert Guido and Berghmans Ron (2000b). In vitro fertilisation: access. responsibility of the doctor and welfare of the child. Unpublished Paper. *Reproductive Ethics Workshop of the TEMPE Project*, Athens, 20-21 October.

Dalla-Vorgia Panagiota and Garanis-Papadatos Tina (2000). The case of Diana. Case Study, *Reproductive Ethics Workshop of the TEMPE Project*, Cork, 16–17 June.

Daniels K. Haimes E. (eds) (1998). *Donor Insemination: International Social Science Perspectives.* Cambridge: Cambridge University Press.

Daniels K. and Lalos O. (1995). The Swedish insemination act and the availability of donors. *Human Reproduction*, **10**: 1871–1874.

Davis Dena S. (1997). Genetic dilemmas and the child's right to an open future. *Hastings Center Report*, **27**(2): 7–15.

Dickenson Donna (1997). *Property, Women and Politics: Subjects or Objects?* Cambridge: Polity Press.

Dickenson Donna (2000). Who Owns Foetal Tissue? Paper, *Reproductive Ethics Workshop of the TEMPE Project*, Cork, 16–17 June. Originally published, in full, in *Ethical Issues in Maternal-Fetal Medicine,* Donna Dickenson, (ed.), Cambridge: Cambridge University Press, 2002, pp. 233–246.

Dondorp Wybo-Jan (2000). Oocytes for research from IVF patients: the role of the doctor. Unpublished Paper, *Reproductive Ethics Workshop of the TEMPE Project*, Athens. 20-21 October.

Donor Conception Group of Australia (2000). Let the Offspring Speak. Donor Conception Group of Australia.

Donor Conception Support Groups (2000). Speaking for Ourselves. Donor Conception Support Groups in the United Kingdom and United States.

Dooley Dolores (2000). Moral Free-Fall in Ethics: Re-thinking Reproductive Responsibility. Upublished Paper, *Reproductive Ethics Workshop of the TEMPE Project*, Athens, 20-21 October.

Dorland William Alexander N. (1994). *Dorland's Illustrated Medical Dictionary,* 28th Edition. Philadelphia: Saunders.

Douglas G. (1994). The intention to be a parent and the making of mothers. *Modern Law Review*, **57**: 636.

Draper Heather and Chadwick Ruth (1999). Beware! Preimplantation genetic diagnosis may solve old problems but it also raises new ones. *Journal of Medical Ethics*, **25**(2): 114–120.

Dyson Anthony and Harris John (eds) (1994). *Ethics and Biotechnology.* London: Routledge.

Edwards J. (1998). Donor insemination and 'public opinion'. In *Donor Insemination: International Social Science Perspectives*, K. Daniels and E. Haimes, (eds), Cambridge: Cambridge University Press, pp. 151–172.

European Group on Ethics and Reproductive Technologies (EGE) (2000). *Ethical Aspects of Human Stell Cell Research and Use.* Opinion No.15. http://europa.eu.int/comm/secretariat general/sgc/ethics/en/index.html (accessed October 2000).

Evans Donald (1990). Legislative control of medical practice. *Bulletin of Medical Ethics*, **55**: 13–17.

Evans Donald (1996). The clinical classification of infertility. In *Creating the Child: the Ethics, Law and Practice of Assisted Reproduction*, Donald Evans, (ed.), The Hague: Martinus Nijhoff Publishers.

Evans M. (1996). A right to procreate? Assisted conception, ordinary procreation and adoption. In *Creating the Child: the Ethics. Law and Practice of Assisted Reproduction*, Donald Evans, (ed.), The Hague: Martinus Nijhoff Publishers.

Fox M. (2000). Pre-persons, commodities or cyborgs: the legal construction and representation of the embryo. *Health Care Analysis*, **8**: 171–188.

Fuchs E. and Segre J. A. (2000). Stem cells: a new lease on life. *Review Cell*, **100**: 143–155.

Garanis-Papadatos Tina (2000a). Understanding Deaf culture: commentary on the case of Linda and Philip. Unpublished Paper, *Reproductive Ethics Workshop of the TEMPE Project*, Athens, 20-21 October.

Garanis-Papadatos Tina (2000b). Pre-implantation genetic diagnosis: commentary on case of Helen and Paul. Unpublished Paper, *Reproductive Ethics Workshop of the TEMPE Project*, Athens, 20-21 October.

Geron Ethics Advisory Board (1999). Research with human embryonic stem cells: ethical considerations. *Hastings Center Report*, **29**(2): 31–38.

Gold E. Richard (1996). *Body Parts: Property Rights and Ownership of Human Biological Materials*. Georgetown: Georgetown University Press.

Goldman Alan H. (1980). *The Moral Foundations of Professional Ethics*. New Jersey: Rowman and Littlefield.

Golombok S., Brewaeys A., Cook R., Giavazzi M. T., Guerra D., Mantovani A., van Hall E., Crosigani P. and Dexues S. (1996). The European study of assisted reproduction families: family functioning and child development. *Human Reproduction*, **11**: 101.

Golombok S., Murray C., Brinsden P. and Abdalla H. (1999). Social versus biological parenting: family functioning and the socioemotional development of children conceived by egg or sperm donation. *Journal of Child Psychology and Psychiatry*, **40**(4): 519–527.

Golombok S., Tasker F. and Murray C. (1997). Children raised by fatherless families from infancy: family relationships and the socioemotional development of children of lesbian and single heterosexual mothers. *Journal of Child Psychology and Psychiatry*, **38**(7): 783–791. (United Kingdom Study)

Gottlieb C., Lalos O. and Lindblad F. (2000). Disclosure of donor insemination to the child: the impact of the Swedish legislation on couples' attitudes. *Human Reproduction*, **15**: 2052–2056.

Grubb Andrew (1998). 'I, me, mine': bodies, parts and property. *Medical Law International*, **3**: 299–313.

Graumann Sigrid (2000a). Commentary on human embryonic stem cells: ethics and policy by Guido de Wert and Ron Berghmans. Unpublished Paper, *Reproductive Ethics Workshop of the TEMPE Project*, Athens, 20-21 October.

Graumann Sigrid (2000b). Prenatal Selection and eugenics: on the possible social consequences of repro-genetics. Unpublished Paper, *Reproductive Ethics Workshop of the TEMPE Project*, Athens, 20-21 October.

Handyside A. and Delhanty J. D. A. (1997). Preimplantation genetic diagnosis: strategies and surprises. *Trends Genetics*, **13**: 270-5.

Harris John (1999). Is gene therapy a form of eugenics? In *Bioethics, An Anthology*, Helga Kuhse and Peter Singer, (eds), London: Blackwell, pp.165–170.

Harris John (2000). Is there a coherent social conception of disability? *Journal of Medical Ethics*, **26** (2): 95–100.

Hastings Center Report (1996). *Special Supplement: The Goals of Medicine*, **26**(6).

Health Council of the Netherlands Committee on *In Vitro* Fertilisation (1995). *Sex Selection for Non-Medical Reasons, 1995/11E*. Rijswijk: Health Council of the Netherlands.

Health Council of the Netherlands Committee on *In Vitro* Fertilisation (1998). *IVF-related Research, 1998/08E*. Rijswijk: Health Council of the Netherlands.

HFE (Human Fertilisation and Embryology) Act (1990). London: HMSO.

Human Fertilisation and Embryology Authority (HFEA) (1998). *Annual Report*. London: HMSO.

Human Fertilisation and Embryology Authority (HFEA) (1998). *Code of Practice* (4th edition). London: HFEA.

Human Genetics Advisory Commission (HGAC) and Human Fertilisation and Embryology Authority (HFEA) (1998). *Cloning Issues in Reproduction. Science and Medicine.* London: Department of Trade and Industry.

Holm S. (1996). Infertility, childlessness and the need for treatment. In *Creating the Child: the Ethics. Law and Practice of Assisted Reproduction*, Donald Evans, (ed.), The Hague: Martinus Nijhoff Publishers.

IFFS Surveillance (1998). *Fertility and Sterility.* Supplement 2, H. W. Jones and J. Cohen, (general eds), **71**(5).

Jerusalem Bible (1985). *Genesis,* Chapter 16. London : Darton, Longman & Todd.

Johnson v Calvert (1993). 851 P.2d 776 Cal.

Joll Nicholas (2001). Intrinsic versus instrumental value. In *The Concise Encyclopedia of the Ethics of New Technologies*, Ruth Chadwick, (ed.), London: Academic Press, pp.267–276.

Kandel R. F. (1994). Which came first: the mother or the egg? A kinship solution to gestational surrogacy. *Rutgers Law Review*, **47**: 165.

Kant Immanuel (1991[1785]). *The Moral Law: Groundwork of the Metaphysics of Morals.* H. J. Paton, (trans.), London: Routledge.

Kant Immanuel (1994[1799]). On a supposed right to lie because of philanthropic concerns. In *Ethical Philosophy*, J. W. Ellington, (trans.), (2nd edition), Cambridge: Hackett Publishing Company.

Kant Immanuel (1994[1797]). Metaphysical principles of virtue. In *Ethical Philosophy*, J. W. Ellington, (trans.), (2nd edition), Cambridge: Hackett Publishing Company.

Kaplan Deborah (1999). Prenatal screening and its impact on persons with disabilities. In *Bioethics: An Anthology*, Helga Kuhse and Peter Singer, (eds.), Oxford: Blackwell Publishers, pp. 130-136.

Klock S. C., Jaco M. C. and Maier D. (1994). A prospective study of donor insemination recipients: secrecy, privacy and disclosure. *Fertility and Sterility*, **62**: 477–484.

Knowles L. P. (1999). Property, progeny and patents. *Hastings Center Report*, **29**(2): 38–40.

Koehn D. (1994). *The Ground of Professional Ethics.* London: Routledge.

Kymlicka Will (1996). Moral philosophy and public policy: the case of new reproductive technologies. In *Philosophical Perspectives on Bioethics*, L.W. Sumner and Joseph Boyle, (eds), Toronto: University of Toronto Press.

Lamb David (1988). *Down the Slippery Slope.* London: Croom Helm.

Lasker J. N. (1998). The users of donor insemination. In *Donor Insemination: International Social Science Perspectives*, K. Daniels and E. Haimes, (eds), Cambridge: Cambridge University Press, pp 7–32.

Lebacqz L. (1985). *Professional Ethics, Power and Paradox.* Nashville: Abingdon Press.

Lee Robert G. and Morgan Derek (1989). Surrogacy: an introductory essay. In *Birthrights: Law and Ethics at the Beginnings of Life*, Robert Lee and Derek Morgan, (eds), London: Routledge, pp.56–60.

Lee Robert G. and Morgan Derek (2001). *Human Fertilisation & Embryology.* London: Blackstone Press, pp. 191–216.

Lunshof J. (2000). Burning issues in Germany. Unpublished Paper, *Ethics and Genetics Workshop of the TEMPE Project*, Maastricht, 9 June.

MacIntyre Alasdair (2001). The two faces of philosophy. Unpublished Paper, *Royal Irish Academy Convocation*, Cork, 18 January.

MacCallum Fiona (2000a). The case of Sarah and John, case and commentary. Unpublished Paper, *Reproductive Ethics Workshop of the TEMPE Project*, Cork, 16–17 June.

MacCallum Fiona (2000b). Commentary on the case of Diana. Unpublished Paper, *Reproductive Ethics Workshop of the TEMPE Project*, Athens, 20-21 October.

MacCallum Fiona and Madden Deirdre (2000). Contract pregnancy. Unpublished Paper, *Reproductive Ethics Workshop of the TEMPE Project*, Cork, 16–17 June.

Madden Deirdre (2000a). The case of baby Emily, case and commentary. Unpublished Paper, *Reproductive Ethics Workshop of the TEMPE Project*, Cork, 16–17 June.

Madden Deirdre (2000b). The case of Molly and Adam Nash, case and commentary. Unpublished Paper, *Reproductive Ethics Workshop of the TEMPE Project*, Athens, 20-21 October.

Madden Deirdre (2000c). The case of the cryopreservation of ovarian tissue: commentary. Unpublished Paper, *Reproductive Ethics Workshop of the TEMPE Project*, Athens, 20-21 October.

Mahowald Mary (1994). As if there were fetuses without women: a remedial essay. In *Reproduction, Ethics and the Law: feminist perspectives*, Joan C. Callahan, (ed.), Bloomington: Indiana University Press, pp. 199–218.

Marcus S. F., Avery S. M., Abusheikha N., Marcus N. K. and Brinsden P. R. (2000). The case for routine HIV screening before IVF treatment, a survey of UK IVF centre policies. *Human Reproduction*, **15** (8): 1657–1661.

Martin J. (1996). Prioritising assisted conception services: a public health perspective. In *Creating the Child: the ethics. law and practice of assisted reproduction*, Donald Evans, (ed.), The Hague: Martinus Nijhoff Publishers.

Mason J. K. and McCall Smith R.A. (1999). *Law and Medical Ethics.* (5th edition) London: Butterworths.

Mason R. O. and Tomlinson G. E. (2001). Genetic Research. In *The Concise Encyclopedia of the Ethics of New Technologies*, Ruth Chadwick, (ed.), London: Academic Press, pp. 165–181.

Mehlman Maxwell and Botkin Jeffrey (1998). *Access to the Genome.* Washington D.C.: Georgetown University Press.

Meilander Gilbert (2001). The point of a ban. *Hastings Center Report*, **31**(1): 9–16.

Mieth Dietmar, Graumann Sigrid and Haker Hille (2000). Preimplantation genetic diagnosis – points to consider. In *The Ethics of Genetics in Human Procreation*, Hille Haker and Deryck Beyleveld, (eds), Aldershot, England: Ashgate, pp. 329–336.

Mill John Stuart (1981 [1859]). *On Liberty*. New York: Penguin.

Moore v. Board of Regents of the University of California (1990). 271 California Reporter 146. Cal. Sup. Ct.

Morgan D. (1985). Making motherhood male: surrogacy and the moral economy of women. *Journal of Law and Society*, **12**: 219.

Nachtigall R. D. (1993). Secrecy: an unresolved issue in the practice of donor insemination. *American Journal of Obstetrics and Gynecology*, **168**: 1846–1851.

Nachtigall R. D., Tschann J. M., Quiroga S. S., Pitcher L. and Becker G. (1997). Stigma, disclosure and family functioning among parents of children conceived through donor insemination. *Fertility and Sterility*, **68**: 83–89.

National Bioethics Advisory Commission (NBAC). (1999). *Ethical Issues in Human Stem Cell Research*, **1**, Maryland: NBAC.

National Institutes of Health (NIH) (1994). *Final Report of the Human Embryo Research Panel*. Bethseda: NIH.

National Institutes of Health (NIH) (2000). *Guidelines for Research Using Human Pluripotent Stem Cells.* www.nih.gov (accessed August).

Nelkin Dorothy and Andrews Lori (1998). Homo economicus: commercialisation of body tissue in the age of biotechnology. *Hastings Center Report.* **28**(5): 30-39.

Nielsen A. F., Pedersen B. and Lauritsen J. G. (1995). Psychosocial aspects of donor insemination, attitudes and opinions of Danish and Swedish donor insemination patients when psychosocial information is being supplied to offspring and relatives. *Acta Obstetrics Gynecology Scandinavia*, **74**: 45–50.

Nuffield Council on Bioethics (1995). *Human Tissue: ethical and legal uses.* London: Nuffield Council on Bioethics.

Nuffield Council on Bioethics (2000). *Stem Cell Therapy: the ethical issues.* London: Nuffield Council on Bioethics.

O' Donnell K. (2000). Legal conceptions: regulating gametes and gamete donation. *Health Care Analysis*, **8**: 137–54.

O' Neill Onora (1991). Kantian ethics. In *A Companion to Ethics*, Peter Singer, (ed.), Oxford: Blackwell, pp. 175–84.

O' Neill Onora (2000). The 'good enough parent' in the age of new reproductive technologies. In *The Ethics of Genetics in Human Reproduction*, Hille Haker and Deryck Beyleveld, (eds)., Aldershot, England: Ashgate, pp. 33–48.

Papp P. (1993). The worm in the bud: secrets between parents and children. In *Secrets in Families and Family Therapy*, E. Imber-Black, (ed.), New York: Norton, pp. 66–85.

Parker Michael (2000). Public deliberation and private choice in genetics and reproduction. *Journal of Medical Ethics*, **26**(3): 160-165.

Parker Michael and Dickenson Donna (2001). *The Cambridge Medical Ethics Workbook*. Cambridge: Cambridge University Press.

Pellegrino E. D. (1983). The healing relationship: the architectonics of clinical medicine. In *The Clinical Encounter: the moral fabric of the patient-physician relationship*, E. E. Shelp, (ed.), Dordrecht: Reidel, pp.153–72.

Pennings G. (1997a). The internal coherence of donor insemination practice: attracting the right type of donor without paying. *Human Reproduction*, **12**: 1842–1844.

Pennings G. (1997b). The 'double track' policy for donor anonymity. *Human Reproduction*, **12**: 2839–2844.

Pennings G. (1999). The welfare of the child. Measuring the welfare of the child: in search of the appropriate evaluation principle. *Human Reproduction*, **14**(5): 1146–50.

Pfefer N. (1993). *The Stork and the Syringe*. Cambridge: Polity Press.

Polkinghorne Committee (1989). *Review of the Guidance on the Research Uses of Fetuses and Fetal Material*. London: HMSO.

Purdy Laura (2000). Surrogate mothering: exploitation or empowerment? In *Bioethics*, Helga Kuhse and Peter Singer, (eds), Oxford: Blackwell, pp. 103 – 112.

Ragoné H. (1994). *Surrogate Motherhood: conception in the heart*. Boulder, Colorado: Westview Press.

Reiss Michael (2001). Biotechnology. In *The Concise Encyclopedia of the Ethics of New Technologies*, Ruth Chadwick, (ed.), London: Academic Press, pp.13–26.

Robertson J. A. (1994). *Children of Choice: freedom and the new reproductive technologies*. Princeton: Princeton University Press.

Royal Commission on New Reproductive Technologies (1993). *Proceed with Care. Final Report*. Ottawa: Communications Group Publishing.

R v Human Fertilisation and Embryology Authority, ex parte Blood (1997). 2 All ER 687 (CA).

R v Kelly (1998). 3 All ER 741. at 749.

Rubenstein D., Thomasma D. C. and Schon E. A. (1995). Germ-line therapy to cure mitochondrial disease: protocol and ethics of *in vitro* ovum nuclear transplantation. *Cambridge Quarterly of Healthcare Ethics*, **4**: 316–9.

Rumball A. and Adair V. (1999). Telling the story: parents scripts for donor offspring. *Human Reproduction*, **14**: 1392–1399.

Schiff A. (1995). Solomonic decisions in egg donation: unscrambling the conundrum of legal maternity. *Iowa Law Review*, **80**: 265.

Schnitter J. J. (1995). *Let me Explain: a book about DI*. Indianapolis: Perspective Press.

Shenfield Francoise (2000). From pre-natal testing to pre-gravid diagnosis: ethical issues in PGD, a question of responsibility. Unpublished Paper, *Reproductive Ethics Workshop of the TEMPE Project*, Cork, 16–17 June.

Shenfield Francoise and Steel S.J. (1997). What are the effects of anonymity and secrecy on the welfare of the child in gamete donation? *Human Reproduction*, **12**: 392–395.

Shultz M. (1994). Legislative regulation of surrogacy and reproductive technology. *University of San Francisco Law Review*, **28**: 613.

Sidaway v. Bethlem Royal Hospital Governors (1985) AC 871 (HL) = K 103.

Silver Lee M. (1999). *Remaking Eden*. London: Orion Books Ltd.

Smith T. (1999*). Ethics in Medical Research, A Handbook of Good Practice*. Cambridge: Cambridge University Press.

Snowden R. and Snowden E. (1998). Families created through donor insemination. In *Donor Insemination: international social science perspectives*, K. Daniels and E. Haimes, (eds), Cambridge: Cambridge University Press, pp 33–52.

Sokolowski R. (1991). The fiduciary relationship and the nature of the professions. In *Ethics, Trust and the Professions*, E. D. Pellegrino, R. M. Veatch and J. P. Langan, (eds), Washington: Georgetown University Press, pp.23–43.

Spallone (1989). *Beyond Conception: the new politics of reproduction*. Massachusetts: Bergin and Garvey.

Sugarman J., Kaalund V., Kodish E., Marshall M. F., Reisner E.G., Wilfond B. S., Wolpe P. R., Capron A. M., Faden R. R., Hage M. L.,

Harvath L., Howe C., Kass N. E., Kurtzberg J., Lederer S., Macklin R., McCullough L. B., Nemo G. J., Olson J., Powers M. and Wallas C. H. (1997). Consensus Statement: ethical issues in umbilical cord blood banking. *JAMA*, **278**(11): 938–943.

Thornton M. H., Francis M. M. and Paulson P. J. (1998). Immature oocyte retrieval: lessons from unstimulated IVF cycles. *Fertility and Sterility*, **70**: 647–50.

Titmuss Richard (1972). *The Gift Relationship: From Human Blood to Special Policy*. New York: Vintage.

Tucker Bonnie Poitras (1998). Deaf culture, cochlear implants and elective disability. *Hastings Center Report*, **28**(4): 6–14.

Turner A. J. and Coyle A. (2000). What does it mean to be a donor offspring? The identity experiences of adults conceived by donor insemination and the implications for counselling and therapy. *Human Reproduction*, **15**: 2041–2051.

Van der Akker O. (2000). The importance of a genetic link in mothers commissioning a surrogate baby in the UK. *Human Reproduction*, **15**(8): 1849–1855.

Veatch R. M. (1981). *A Theory of Medical Ethics*. New York: Basic Books.

Veatch R. M. (1988). Models for ethical medicine in a revolutionary age. In *Ethical Issues in Professional Life*, J. Callahan, (ed.), Oxford: Oxford University Press, pp. 89–91.

Waldron Jeremy (1988). *The Right to Private Property*. Oxford: Clarendon Press.

Warnock Committee (1984). *Report of Inquiry Into Human Fertilisation and Embryology*. London: HMSO.

Whittal Hugh (2000). Decision-making in ART: art or science? Upublished Paper, *Reproductive Ethics Workshop of the TEMPE Project*, Cork, 16–17 June.

Widdows Heather (2000). Secrecy in donor insemination. Paper, *Reproductive Ethics Workshop of the TEMPE Project*, Athens, 20–21 October. Originally published, in full, in *Ethical Issues in Maternal-Fetal Medicine*, Donna Dickenson, (ed.), Cambridge: Cambridge University Press, 2002, pp.167–180.

Yeo Michael and Moorhouse Anne (eds) (1996). *Concepts and Cases in Nursing Ethics*. Ontario, Canada: Broadview Press.

Index